Spatial Audio

Music
TECHNOLOGY
S e r i e s

Titles in the series

Spatial Audio

Francis Rumsey

Focal Press
Taylor & Francis Group

NEW YORK AND LONDON

First published 2001

This edition published 2013
by Focal Press
70 Blanchard Road, Suite 402, Burlington, MA 01803

Simultaneously published in the UK
by Focal Press
2 Park Square, Milton Park, Abingdon, Oxon OX14 4RN

Focal Press is an imprint of the Taylor & Francis Group, an informa business

Notices

Practitioners and researchers must always rely on their own experience and knowledge
in evaluating and using any information, methods, compounds, or experiments described
herein. In using such information or methods they should be mindful of their own safety
and the safety of others, including parties for whom they have a professional responsibility.

To the fullest extent of the law, neither the Publisher nor the authors, contributors, or
editors, assume any liability for any injury and/or damage to persons or property as a matter
of products liability, negligence or otherwise, or from any use or operation of any methods,
products, instructions, or ideas contained in the material herein.

British Library Cataloguing in Publication Data
A catalogue record for this book is available from the British Library

Library of Congress Cataloguing in Publication Data
A catalogue record for this book is available from the Library of Congress

ISBN 13: 978-0-240-51623-3 (pbk)

Composition by Scribe Design, Gillingham, Kent, UK

Contents

Series introduction

The Focal Press Music Technology Series is intended to fill a growing need for authoritative books to support college and university courses in music technology, sound recording, multimedia and their related fields. The books will also be of value to professionals already working in these areas and who want either to update their knowledge or to familiarise themselves with topics that have not been part of their mainstream occupations.

Information technology and digital systems are now widely used in the production of sound and in the composition of music for a wide range of end uses. Those working in these fields need to understand the principles of sound, musical acoustics, sound synthesis, digital audio, video and computer systems. This is a tall order, but people with this breadth of knowledge are increasingly sought after by employers. The series will explain the technology and techniques in a manner which is both readable and factually concise, avoiding the chattiness, informality and technical woolliness of many books on music technology. The authors are all experts in their fields and many come from teaching and research backgrounds.

Dr Francis Rumsey
Series Editor

Preface

Since the early part of the twentieth century, perhaps even the last years of the nineteenth, sound engineers have been aware of the need for a spatial dimension to their art. For reasons mainly to do with commercial feasibility, a lot of mainstream approaches to this issue have been limited to only two audio channels, feeding two loudspeakers intended to be placed in front of the listener. This resulted in the need for a number of compromises and limited the creative options for rendering spatial sound. Although cinema sound has involved more than two channels for many years, most consumer audio has remained resolutely two-channel.

The later part of the twentieth century, particularly the last ten years, gave rise to a rapid growth in systems and techniques designed to enhance the spatial quality of reproduced sound, particularly for consumer applications. Larger numbers of loudspeakers became common and systems capable of rendering fully three-dimensional sound images were realised by means of the digital signal processing power available in relatively low-cost products. At the beginning of the twenty-first century, sound engineers outside the movie industry are at last in the fortunate position of being able to break free of the limitations of conventional two-channel stereo, either using binaural signal processing techniques or multichannel loudspeaker reproduction. The recent surround sound developments in consumer entertainment and cinema sound systems, whilst clearly another commercial compromise like two channel stereo, offer greater creative freedom to manipulate sound in the spatial domain, and to create immersive sound fields for music, games, television and other applications.

The spatial aspect of sound quality has remained largely unimproved for many years in mainstream consumer audio applications. While it is acknowledged that numerous research projects and experimental approaches have taken place over the years, it is only recently that a major step change has taken place in the equipment that consumers can use to replay sound in home entertainment systems. Quadraphonic sound did not succeed commercially in the 1970s and binaural audio remained largely a fascinating research tool for many years. Ambisonics, again an elegant concept, failed to capture the commercial imagination. Today it is increasingly likely that one will encounter multichannel surround sound systems in the home (driven strongly by the 'home cinema' revolution), or that one will find binaural technology implemented in consumer products such as televisions and computer sound systems. The means of delivering more than two channels of audio to the consumer now exist in the form of DVD and other digital media services, and DSP has now reached the point that the complex filtering and manipulation of sound signals necessary to process signals binaurally can be achieved in low-cost products.

Reproduced sound quality, in the sense that most people understand it, has improved dramatically in the past forty years. From relatively modest signal-to-noise ratio, distortion performance and frequency response we have arrived at a point where these things are simply not a problematic issue for most people. The *technical* quality of sound recording systems can now be made arguably as good as it needs to be to capture the dynamic and frequency ranges that we can perceive. *Spatial* sound quality, then, is possibly the only major factor remaining to be tackled in the quest for ultimate quality in sound reproduction. Thanks to recent enabling developments we can take some more steps towards this goal.

The aims of spatial sound reproduction, and how they can be achieved by technical means, are the main subject of this book. It requires that we look into issues of sound perception as well as the technology and techniques that can be adopted. A strong emphasis is placed on the acceptance of recent standard configurations of loudspeakers. Although the '5.1-channel' approach to surround sound reproduction is not ideal, it is a compromise that took many years to hammer out, representing a means of maintaining compatibility with two-channel stereo, with surround sound in the cinema and with television applications, whilst offering a real possibility for spatial enhancement. Many have argued that it is not the best option for music reproduction, and while this argument has some merit it must be accepted that the chance of people installing two different systems in their homes (one for watching movies and the other for listening to music) is very small in most cases.

The rise of surround sound systems, either using multiple loudspeakers or 'virtual' (binaurally generated) sources using fewer transducers, presents an enormous challenge to sound engineers, many of whom are confused by the possibilities and unfamiliar with standards, formats, track allocations, monitoring configurations and recording techniques. Although this book does not pretend to offer all the answers to these people, it is intended as a comprehensive study of the current state of the art in spatial audio.

In writing this book I have relied upon the excellent work of a wide range of people working in spatial audio, too numerous to mention. My own research students and colleagues have also helped me to understand many of the issues contained herein and have undertaken valuable experimental work. I am particularly grateful to the other members of the Eureka MEDUSA project, a group that enabled me to learn a great deal more about spatial audio and psychoacoustics. The AES Technical Committee on Multichannel and Binaural Audio has proved to be an excellent forum for debate and information. David Griesinger is to be thanked for a number of inspiring discussions, papers and lectures which caused me to question what I thought I knew about this subject. Michael Gerzon was always patient in explaining complicated issues and never dismissive. Here was a person who, despite his considerable intellect, was always willing to spend as long as it took to explain a matter at an appropriate level without making his interlocutor feel stupid. Every time I go back to his papers I realise how far ahead of his time he was. Finally, I am indebted to my long-time colleague and friend, Dave Fisher, for his detailed comments on the manuscript. Responsibility for the final product is, of course, mine.

Francis Rumsey

1 Introduction to spatial audio

1.1 The spatial dimension in natural sound

Everyday life is full of three-dimensional sound experiences. The ability of humans to make sense of their environments and to interact with them depends strongly on spatial awareness and hearing plays a major part in this process. Natural sounds are perceived in terms of their location and, possibly less consciously, their size (most people rely more strongly on their visual sense for this information). Typically natural sound environments contain cues in all three dimensions (width, height, depth) and one is used to experiencing sounds coming from all around, with no one direction having particular precedence over any other. Because listeners don't have eyes in the backs or tops of their heads, they tend to rely on vision more to perceive the scene in front of them, and on sound more to deal with things behind and above them.

Most naturally experienced sound environments or sound 'scenes' consist of numerous sources, each with its own location and attributes. Complex sound scenes may result in some blending or merging of sounds, making it more difficult to distinguish between elements and resulting in some grouping of sources. In some cases the blending of cues and the diffuseness of sources leads to a general perception of space or 'spaciousness' without a strong notion of direction or 'locatedness' (as Blauert calls it) (Blauert, 1997). One only has to imagine the sound of an outdoor environment in the middle of the country to appreciate this:

there is often a strong sense of 'outdoorness' which is made up of wind noise, general background noise from distant roads and towns, punctuated with specific localisable sounds such as birds. This sense of 'outdoorness' is strongly spatial in character in that it is open rather than constricted, and is very much perceived as outside the listener's head. Outdoor environments are not strongly populated with sound reflections as a rule, compared with the sound experienced inside rooms, and those reflections that one does experience are often quite long delayed (apart from the one off the ground). Outdoor sources can be very distant compared with indoor sources.

Indoor environments have surfaces enclosing the sound sources within them, so tend to be strongly affected by the effect of reflections. One's spatial sense in a room attempts to assess the size of the space, and the distance of objects tends to be within a few tens of metres. In many places in many rooms one's perception of sound is strongly dominated by the reflected sound, so the reflections play a large part in the spatial characteristics of sources and tend to modify them. The reflections in most real rooms tend to be within a relatively short time after the direct sound from the source.

Overall then, the spatial characteristics of natural sounds tend to split into 'source' and 'environment' categories, sources being relatively discrete, localisable entities, and environments often consisting of more general 'ambient' sound that is not easily localised and has a diffuse character. Such ambient sound is often said to create a sense of envelopment or spaciousness that is not tied to any specific sound source, but is a result of reflections, particularly in indoor environments. The spaciousness previously referred to as 'outdoorness' is much less related to reflections, probably being more strongly related to the blending of distant sound sources that have become quite diffuse. Later on we will consider more detailed issues of psychoacoustics related to this topic.

1.2 Sound sources in space

1.2.1 Sound sources in a free field

In acoustics, the free field is a term used to describe an environment in which there are no reflections. The closest most people might get to experiencing free field conditions is outdoors, possibly suspended a long way above the ground and some way from buildings (try bungee jumping or hang-gliding). Those involved in acoustics research might have access to an anechoic chamber,

Figure 1.1 Change in intensity of direct sound with increasing distance from a source. (a) An omnidirectional source radiates spherically. (b) Sound energy that passed through 1 m² of the sphere's surface at distance *r* will have expanded to cover 4 m² at distance 2*r*, giving rise to one quarter of the intensity or a 6 dB drop.

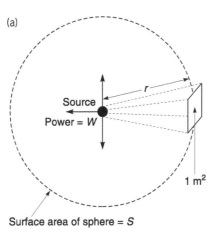

(a)

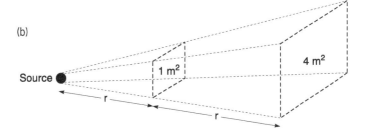

(b)

which is an artificial environment created within a large room containing copious amounts of shaped absorbing material that reduces reflections to a minimum, creating a virtual free field over the majority of the audio frequency range. Such environments are rarely experienced in normal life, making them quite uncanny when they are experienced.

In the free field, all the sound generated by a source is radiated away from the source and none is reflected back. A consequence of this is that the sound level experienced by a listener drops off quite rapidly as they move away from the source (about 6 dB for every doubling in distance from the source) as shown in Figure 1.1. This is because the sound energy is distributed over a sphere of ever-increasing surface area as it expands away from the source. So close to a source the level will drop quite quickly as one moves away from it, gradually dropping less quickly as one gets further away. At some distance from the source the wave front curvature becomes so shallow that the wave can be considered a plane wave, for most purposes.

Sounds are relatively easy to localise in free field environments as the confusing effect of reflections is not present. As mentioned

Figure 1.2 The directivity pattern of a source shows the magnitude of its radiation at different angles. The source shown here has a pattern that is biased towards the front (0° axis), and increasingly so at higher frequencies.

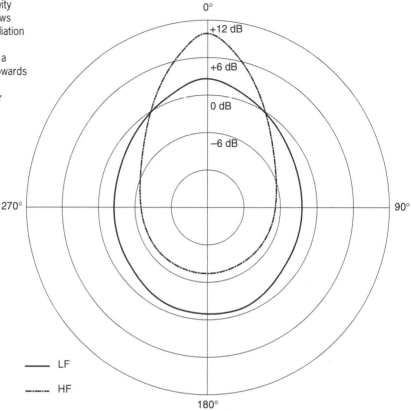

later, the human listener localises a source by a combination of time and level/spectrum difference measurement between the ears. Distance or depth judgement is not so straightforward in free field environments, as we shall see, because all one really has to go on is the loudness of the source (and possibly the high frequency content if it is a long way away), making absolute judgements of distance quite difficult.

Not all sources radiate sound spherically or omnidirectionally, indeed most have a certain directivity characteristic that represents the deviation from omnidirectional radiation at different frequencies. This is sometimes expressed as a number of dB gain compared with the omnidirectional radiation at a certain frequency on a certain axis (usually the forward or 0° axis). This is best expressed fully as a polar pattern or directivity pattern, showing the directional characteristics of a source at all angles and a number of frequencies (see Figure 1.2). As a rule, sources tend to radiate more directionally as the frequency rises, whereas low frequency radiation is often quite omnidirectional

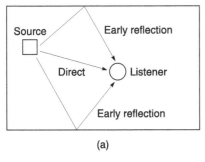

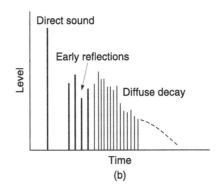

(a)

(b)

Figure 1.3 The response of an enclosed space to a single sound impulse. (a) The direct path from source to listener is the shortest, followed by early reflections from the nearest surfaces. (b) The impulse response in the time domain shows the direct sound, followed by some discretely identifiable early reflections, followed by a gradually more dense reverberant tail that decays exponentially.

(this depends on the size of the object). This can be quite important in the understanding of how microphones and loudspeakers interact with rooms, and how sound sources will be perceived spatially.

1.2.2 Sources in reflective spaces

In enclosed spaces a proportion of the radiated sound energy from sources is absorbed by the surfaces and air within the space and a proportion is reflected back into the environment. The result of the reflected sound is to create, after a short period, an 'ambient' or 'diffuse' sound field that is the consequence of numerous reflections that have themselves been reflected. As shown in Figure 1.3, the response of a space to a short sound impulse is a series of relatively discrete early reflections from the first surfaces encountered, followed by a gradually more dense and diffuse reverberant 'tail' that decays to silence.

In reflective spaces the sound level does not drop off as rapidly as one moves away from a sound source because the reflected sound builds up to create a relatively unchanging level of diffuse sound throughout the space, as shown in Figure 1.4. Although the direct sound from a source tails off with distance in the same way as it would in the free field, the reflected sound gradually takes over. At some distance, known as the critical distance or room radius, the direct and reflected sound components are equal in level. Beyond this reflected sound dominates. This distance depends on the level of reflected sound in the room which is in turn related to the room's reverberation time (the time taken for a sound's reverberation to decay to a level 60 dB below the source's original level). The critical distance can be calculated quite easily if one knows a few facts about the room and the source:

Critical distance = $0.141\sqrt{RD}$

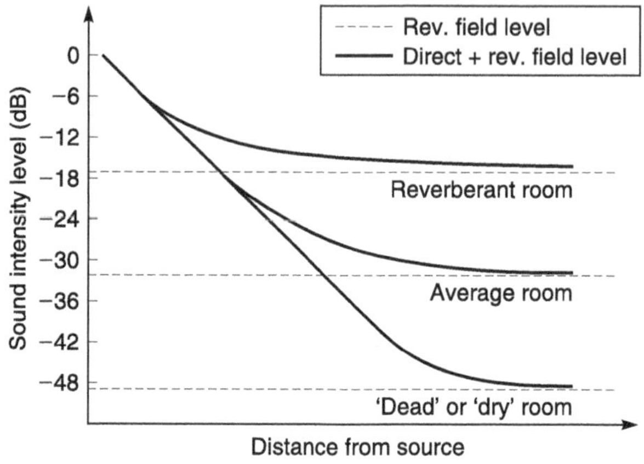

Figure 1.4 As the distance from a source increases, direct sound level falls but reverberant sound level remains roughly constant. The resulting sound level experienced at different distances from the source depends on the reverberation time of the room, because the level of reflected sound is higher in a reverberant room than in a 'dead' room. The critical distance is the point at which direct and reverberant components are equal in each case.

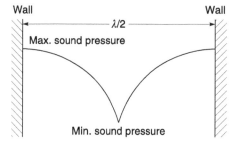

Figure 1.5 Pressure pattern of the first standing wave mode resulting between two room surfaces (called an axial mode). The first mode occurs when the distance between the surfaces equals half the sound wavelength. Further modes occur at multiples of such frequencies, and also for paths involving four or six surfaces (tangential and oblique modes).

where R is the so-called 'room constant' and is related to the rate of absorbtion of sound in the space. $R = S\alpha/(1-\alpha)$ where S is the total surface area of the room in square metres and α is the average absorbtion coefficient of the room. D is the directivity factor of the source (equal to 1 for an omnidirectional source). Directivity factor is the ratio of the sound pressure level on the front or normal axis of radiation to that which would be observed with an omnidirectional source.

The effect of reflections is also to create so-called room modes or *eigentones* at low frequencies, which are patterns of high and low sound pressure resulting from standing waves set up in various combinations of dimensions. These modes occur at frequencies dictated by the dimensions of a space, as shown in Figure 1.5. Sources can be either strongly coupled to these modes or weakly coupled, depending on whether they are located near antinodes or nodes of the mode (pressure maxima or minima). If they are strongly coupled they will tend to excite the mode more than when they are weakly coupled.

1.2.3 Introduction to effects of reflections and reverberation

Reflections have the effect of modifying the perceived nature of discrete sound sources. The early reflections have been found to contribute strongly to one's sense of the size and space of a room, although in fact they are perceptually fused with the direct sound in most cases (in other words they are not perceived as discrete echoes). Early reflections have been found to affect one's perception of the size of a sound source, while slightly later reflections have been found to contribute more to a sense of spaciousness or envelopment (this is discussed in more detail in Chapter 2).

Localisation of sound sources can be made more difficult in reflective environments, although the brain has a remarkable ability to extract useful information about source location from reverberant signals. Distance or depth perception has often been found to be easier in reverberant spaces because the timing of the reflected sound provides numerous clues to the location of a source, and the proportion of reflected to direct sound varies with distance. Also the boundaries and therefore maximum distances in a room are typically strongly established by the visual sense.

Reflections in sound listening environments have been shown to have an important effect on the timbral and spatial qualities of reproduced sound, leading to a variety of designs for sound mixing and monitoring environments that attempt to control the level and timing of such reflections. This is covered in Chapter 5.

When considering recording techniques it is useful to appreciate the issue of critical distance in a room. When microphones are closer than this to the source one will pick up mainly direct sound, and when further away mainly reverberant sound. It is also apparent that there is a relationship between critical distance and decorrelation of reverberant sound components in a room, which in turn has been related to the important attribute of spaciousness in sound recordings and recommendations for microphone spacing. More of this in Chapters 6 and 7.

1.3 Introduction to the spatial dimension in reproduced sound

1.3.1 What is the aim of sound reproduction?

Arguments have run for many years surrounding the fundamental aesthetic aim of recording and reproducing sound. In classical music recording and other recording genres where a natural environment is implied or where a live event is being relayed it is often said that the aim of high quality recording and reproduction should be to create as believable an illusion of 'being there' as possible. This implies fidelity in terms of technical quality of reproduction, and also fidelity in terms of spatial quality. Others have suggested that the majority of reproduced sound should be considered as a different experience from natural listening, and that to aim for accurate reconstruction of a natural sound field is missing the point – consumer entertainment in the home being the aim. Many commercial releases of music exemplify this latter view. Some have likened the dilemma facing sound engineers to the difference between a 'you are there' and a 'they are here' approach to sound recording – in other words, whether one is placing the listener in the concert hall environment or bringing the musicians into his living room.

This, of course, completely ignores the large number of music recordings and cinema releases where there is no natural environment to imply or recreate. In many cases of pop music and cinema or TV sound one is dealing with an entirely artificial creation that has no 'natural' reference point or perceptual anchor. Here it is hard to arrive at any clear paradigm for spatial reproduction, as the acoustic environment implied by the recording engineer and producer is a form of 'acoustic fiction'. (This argument is developed further at the start of Chapter 6.) Nonetheless, it would not be unreasonable to propose that spatial experiences that challenge or contradict natural experience (or that are suggested by the visual sense) might lead to discomfort or dissatisfaction with the product. This is not to deny the possibility of using challenging or contradictory spatial elements in reproduced sound for intentional artistic effect – indeed many artists have regarded the challenging of accepted norms as their primary aim – but more to suggest the need for awareness of these issues in sound balancing. North and Hargreaves (1997) suggest that there is an optimum degree of complexity in music that leads to positive listener responses ('liking'), and this might well be expected to extend to the spatial complexity of reproduced music. Figure 1.6 shows the typical arched curve that comes from these studies, suggesting that beyond a certain point increased complexity in music results in a drop off in liking. Those balancing surround sound recordings of pop music have certainly found that great care is needed in the degree to which sources are placed behind the listener, move about and generally confuse the overall 'picture' created. These issues are covered in more detail in Chapters 6 and 7.

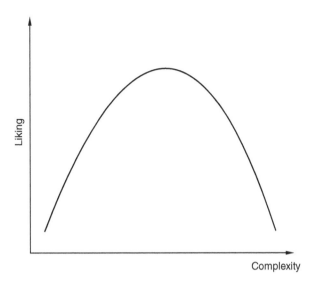

1.3.2 Spatial attributes in sound reproduction

Considerable research has been conducted in the past few years, both at the author's Institute and elsewhere, to determine what people perceive when listening to reproduced sound and how that perception relates to their preference and to the physical variables in sound systems. This is covered further in Chapter 2, but here it is simply noted that while many people have tended to concentrate in the past on analysing the ability of spatial sound systems to create optimally localised phantom images and to reconstruct original wavefronts accurately, recent evidence suggests that other (possibly higher level) subjective factors such as image depth, width and envelopment relate strongly to subjective preference.

If true identity were possible (or indeed desirable) between recording environment and reproducing environment, in all three dimensions and for all listening positions, then it might be reasonable to suppose that ability of the reproducing system to create accurate phantom images of all sources (including reflections) would be the only requirement for fidelity. Since true identity is rarely possible or desirable we return to the notion that some means of creating and controlling adequate illusions of the most important subjective cues should be the primary aim of recording and reproducing techniques. To those with the view that accurate localisation performance is the only true factor of importance in assessing the spatial performance of a sound reproducing system one might ask how they know that this would also achieve accurate reproduction of factors such as spaciousness and envelopment. The latter factors are much harder to define and measure, but they appear nonetheless to be quite important determinants of overall quality.

1.4 From mono to surround sound and 3D audio – a brief resumé

A short summary of some of the major developments in spatial audio systems is given here. Many of the details are reserved for later chapters.

1.4.1 Early sound reproducing equipment

The first gramophone and phonograph recording systems from the late 1800s and early 1900s were monophonic (one channel only). (There was a remarkable three-horned device called the 'Multiplex Graphophone Grand' from Columbia at one point, playing back three separate grooves from the recording medium, but it was not believed to be inherently a stereo recording system, although there might have been some limited phase differences between the signals.) The only 'spatial' cues possible in monophonic reproduction were hints at distance and depth provided by reverberation.

1.4.2 The first stereo transmission?

Clement Ader's early experiment at the Paris exhibition of 1881 is often documented as the first known example of a stereophonic transmission of music (Hertz, 1981). He placed telephone pickups (microphones) in the footlights at the Paris Opera (spaced across the stage) and relayed the outputs of these to pairs of telephone receiver earpieces at the exhibition, where delighted visitors could listen to the opera live and with some spatial realism. Unfortunately it was not until many years afterwards that stereophonic reproduction became a commercial reality.

1.4.3 Bell Labs in the 1930s

Early work on directional reproduction at Bell Labs in the 1930s involved attempts to approximate the sound wavefront that would result from an infinite number of microphone/loudspeaker channels by using a smaller number of channels, as shown in Figure 1.7. Spaced pressure (omnidirectional) microphones were used, each connected by a single amplifier to the appropriate loudspeaker in a listening room. Steinberg and

Figure 1.7 Steinberg and Snow's attempt to reduce the number of channels needed to convey a source wavefront to a reproduction environment with appropriate spatial features intact. (a) 'Ideal' arrangement involving a large number of transducers. (b) Compromise arrangement involving only three channels, relying more on the precedence effect.

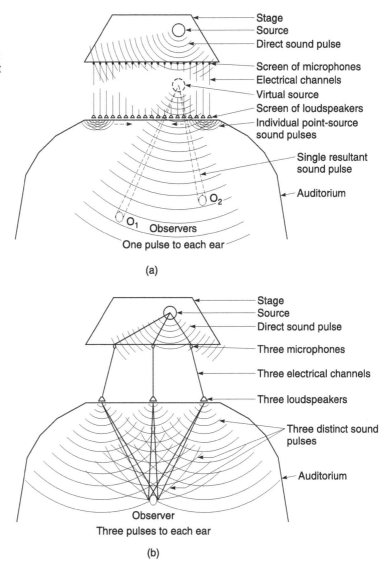

Stage
Source
Direct sound pulse
Screen of microphones
Electrical channels
Virtual source
Screen of loudspeakers
Individual point-source sound pulses
Single resultant sound pulse
Auditorium
O_2
O_1 Observers
One pulse to each ear

(a)

Stage
Source
Direct sound pulse
Three microphones
Three electrical channels
Three loudspeakers
Three distinct sound pulses
Auditorium
Observer
Three pulses to each ear

(b)

Snow (1934) found that three channels gave quite convincing results, and that when reducing the number of channels from three to two, central sources appeared to recede towards the rear of the stage and that the width of the reproduced sound stage appeared to be increased. In fact, as Snow later explained, the situations shown in the two diagrams are in fact rather different because the small number of channels do not really recreate the original source wavefront, but depend upon the precedence effect (see Chapter 2) for success.

Steinberg and Snow's work was principally intended for large auditorium sound reproduction with wide screen pictures, rather than small rooms or consumer equipment. It is interesting to note that three front channels, although not used much in consumer reproduction until recently, are the norm in cinema sound reproduction, partly because of the wide range of seating positions and size of the image. The centre channel has the effect of stabilising the important central image for off-centre listeners, and has been used increasingly since the Disney film *Fantasia* in 1939 (*Fantasia* used a special control track to automate the panning of three sound tracks to a number of loudspeakers, including some rear channels).

1.4.4 Blumlein's patent

The difference between binaural perception of a single sound (in which a single source wavefront is heard separately by the two ears, as in natural listening) and what Snow called 'stereophonic situations' (in which multiple loudspeaker signals are used to create 'phantom images') was recognised by Alan Blumlein, whose now famous patent specification of 1931 (Blumlein, 1931) allows for the conversion of signals from spaced pressure microphones (generating phase differences relating to the source position) to a format suitable for reproduction on loudspeakers. He showed that by introducing only amplitude differences between a pair of loudspeakers it would be possible to create phase differences between the ears, similar to those in natural listening. The patent also allows for other formats of pickup that result in an approximation of the original source phase differences at the ears when reproduced on loudspeakers, leading to the now common coincident-pair microphone techniques. Blumlein's was a system that was mainly designed to deal with source imaging accuracy over a limited angle, and mainly at low frequencies, although it also conveyed other spatial attributes reasonably well.

Blumlein's work remained unimplemented in commercial products for many tens of years, and much writing on stereo reproduction, even in the 1950s, appears unaware of his work. A British paper presented by Clark, Dutton and Vanderlyn of EMI in 1958 revived the Blumlein theories, and showed in more rigorous detail how a two loudspeaker system might be used to create an accurate relationship between the original position of a source and the perceived position on reproduction, by controlling only the relative signal amplitudes between two loudspeakers (derived in their case from a pair of coincident figure-eight microphones).

The authors discuss the three channel spaced microphone system of Bell Labs and suggest that although it produces convincing results in many listening situations it is uneconomical for domestic use, and that the two channel simplification (using two spaced microphones about 10 feet apart) has a tendency to result in a subjective 'hole in the middle' effect (an effect with which many modern users of spaced microphones may be familiar, in which the sound appears to come from either left or right, but without a clear centre image). They concede in a discussion that the Blumlein method adapted by them does not take advantage of all the mechanisms of binaural hearing, especially the precedence effect, but that they have endeavoured to recreate a few of the directional cues that exist in the natural listening.

1.4.5 Early consumer stereo

Methods for cutting two channel stereo sound onto vinyl disks, in a manner similar to that proposed by Blumlein, were introduced commercially in the late 1950s, and stereo records became widely available to the public in the 1960s. Quite a lot of recordings were issued as 'monophonic recording artificially reprocessed to give stereophonic effect on stereophonic equipment' (using various comb filter and band splitting techniques to distribute the sound energy between channels, with varying degrees of success). The Beatles' album, *Sergeant Pepper*, from the mid '60s, was the first multitrack pop recording (four track!) and was issued in both stereo and mono versions (the stereo being the more common, but some say the poorer mix). Early pop stereo was often quite crude in its directional effects, leading to the coining of the term 'ping-pong stereo' to describe the crude left–right division of the instruments in the mix. Also in the 1960s, FM radio with stereo capability was launched, providing a level of quality not previously experienced by the majority of consumers.

1.4.6 Binaural stereo

It could well be argued that all sound reproduction is ultimately binaural because it is auditioned by the two ears of a listener. Nonetheless the term binaural stereo is usually reserved for signals that have been recorded or processed to represent the amplitude and timing characteristics of the sound pressures present at two human ears.

The method of recording stereo sound by using two microphones located in the ears of a real or dummy head has been popular in academic circles for a long time, owing to its potential ability to encode all of the spatial cues received by human listeners. Documented examples of interest in this approach go back over much of the twentieth century. When reproduced over headphones such recordings can recreate a remarkable sense of realism, including height cues and front–back discrimination. Unfortunately the variables in such signal chains, particularly the differences between the recording head/ears and the listener's, the headphone response and coupling to the ears, and any distortions in the signal path, can easily destroy the subtle spectral and timing cues required for success. Furthermore, the important effect of head movements that enable natural listeners to resolve front–back confusions and other localisation errors is not present as a rule with binaural reproduction.

Binaural recordings are not immediately compatible with loudspeaker listening, resulting in a potential barrier to widespread commercial use, although they can be processed to be so. Also, although a simple dummy head ought theoretically to be the most *accurate* way of recording music for headphone reproduction, much recorded music is artificially balanced from multiple microphones and the factors differentiating commercial sound balances from natural listening experiences (mentioned earlier) come into play. This is covered further in Chapter 3.

Recent developments in digital signal processing (1990s) have resulted in a resurgence of interest in binaural technology, often going under titles such as '3D audio' and 'virtual surround'. It is now possible to process multiple tracks of audio to mix sources and pan them binaurally, using digital representations of the auditory responses involved. Binaural material can be processed more easily for reproduction on loudspeakers, and various systems are in wide use in computer sound cards and consumer televisions for spatial enhancement of the sound from only two loudspeakers. It is nonetheless normal to expect such systems only to work satisfactorily for a very limited range of listening positions. Virtual reality systems and computer games environments benefit considerably from such enhancements, and head tracking is increasingly used to incorporate head movements into the processing equation.

1.4.7 Cinema stereo

Apart from the unusual stereo effects used in *Fantasia* (as mentioned above), cinema sound did not incorporate stereo reproduction until the 1950s. Stereo film sound tracks often employed dialogue panned to match the visual scene elements, which was a laborious and time consuming process. This technique gradually died out in favour of central dialogue, accompanied by stereo music and sound effects. During the 1950s

Warner Brothers introduced a large screen format with three front channels and a single surround channel, and the 20th Century Fox *Cinemascope* format also used a similar arrangement.

Multichannel stereo formats for the cinema became increasingly popular in the late '50s and 1960s, culminating in the so-called 'baby boomer' 70 mm format involving multiple front channels, a surround channel and a subwoofer channel to accompany high quality, wide-screen cinema productions. In the early '70s, Dolby's introduction of Dolby Stereo enabled a four channel surround sound signal to be matrix encoded into two optical sound tracks recorded on the same 35 mm film as the picture, and this is still the basis of the majority of analogue matrix film sound tracks today, having been released in a consumer form called Dolby Surround for home cinema applications. The main problem with analogue matrix formats was the difficulty of maintaining adequate channel separation, requiring sophisticated 'steering' circuits in the decoder to direct dominant signal components to the appropriate loudspeakers.

Modern cinema sound is gradually moving over to all-digital sound tracks that typically incorporate either five or seven discrete channels of surround sound plus a sub-bass effects channel. A variety of commercial digital low-bit-rate coding schemes are used to deliver surround sound signals with movie films, such as Dolby Digital, Sony SDDS and Digital Theatre Systems (DTS), as described in Chapter 4.

1.4.8 Ambiophony and similar techniques

Although surround sound did not appear to be commercially feasible for consumer music reproduction applications during the late 1950s and early 1960s, a number of researchers were experimenting at the time with methods for augmenting conventional reproduction by radiating reverberation signals from separate loudspeakers. This is an interesting precursor of the modern approach that tends to recommend the use of surround channels for the augmentation of conventional frontal stereo with ambience or effects signals. One of the most interesting examples in this respect was the 'Ambiophonic' concept developed by Keibs and colleagues in 1960.

1.4.9 Quadraphonic sound

Quadraphonic sound is remembered with mixed feelings by many in the industry, as it represents a failed attempt to introduce surround sound to the consumer. A variety of competing encoding methods, having different degrees of compatibility with each other and with two channel stereo, were used to convey four channels of surround sound on two channel analogue media such as vinyl LPs (so-called 4–2–4 matrix systems). Unlike Dolby Stereo, quadraphonic sound used no centre channel, but was normally configured for a square arrangement of loudspeakers, two at the front and two behind the listener. The 90° angle of the front loudspeakers proved problematic because of lack of compatibility with ideal two channel reproduction, and gave poor front images, often with a hole in the middle.

While a number of LP records were issued in various quad formats, the approach failed to capture a sufficiently large part of the consumer imagination to succeed. It seemed that people were unwilling to install the additional loudspeakers required, and there were too many alternative forms of quad encoding for a clear 'standard' to emerge. Also, many people felt that quad encoding compromised the integrity of two channel stereo listening (the matrix encoding of the rear channels was supposed to be two-channel compatible but unwanted side effects could often be heard).

1.4.10 Ambisonics

Ambisonic sound was developed in the 1970s by a number of people including Gerzon, Fellgett and Barton, and many others influenced its development both in the early days and since. Much of the work was supported by the NRDC (National Research and Development Council) and the intellectual property was subsequently managed by the British Technology Group (this was eventually transferred to the British record company, Nimbus). It was intended as a comprehensive approach to directional sound reproduction, involving any number of reproduction channels, based partly on an extension of the Blumlein principle to a larger number of channels. The system can be adapted for a wide variety of loudspeaker arrangements, including (more recently) the ITU-standard five-channel configuration. Enabling fully three-dimensional sound fields to be represented in an efficient form, including a height component, it remains an elegant technical toolbox for the sound engineer who believes that accurate localisation vector reconstruction at the listening position is the key to high quality spatial sound reproduction.

Ambisonics is discussed in greater detail in Chapter 4, but here it is simply noted that despite its elegance the technology did not gain ready acceptance in the commercial field. Although a number of consumer Ambisonic decoders and professional devices were manufactured, there are few recordings available in the format (a matrixing method known as UHJ enables Ambisonic surround recordings to be released on two channel media). This has sometimes been attributed to the somewhat arcane nature of the technology, Ambisonics being mainly a collection of principles and signal representation forms rather than a particular implementation. The apparent difficulty of grasping and marketing Ambisonics as a licensable entity, or indeed of communicating the potential advantages, seems to have led to it languishing among a small band of dedicated enthusiasts who still hope that one day the rest of the world will see sense. Nonetheless it has found increasing favour as an internal representation format for sound fields in recent 3D audio products, and may yet become more widely used than it has been to date.

1.4.11 The home cinema and ITU-standard surround sound

In recent years the development of new consumer audio formats such as DVD, and digital sound formats for cinema and broadcasting such as Dolby Digital and DTS, have given a new impetus to surround sound. The concept of the home cinema has apparently captured the consumer imagination, leading to widespread installation of surround sound equipment in domestic environments. This trend appears to be increasing, and it seems likely that surround sound will find its way into the home as a way of enhancing the movie-watching experience (people have become used to surround sound in the cinema and want it at home). Where music reproduction alone appeared to be insufficient justification for reconfiguring the furniture in the living room, movie watching is regarded as an enjoyable experience for all the family, removing some of the traditional barriers to the installation of hi-fi equipment. Where quadraphonics failed to win round the market, home cinema is succeeding. The reason for this is that all the right conditions are in place at the same time – increased spending on entertainment and leisure compared with 30 years ago, pictures, technical quality, a truly mass market and digital delivery media.

Recent digital formats typically conform to the ITU 5.1-channel configuration (three front channels, two surround channels and an optional sub-bass effects channel that is known as the '0.1' channel owing to its limited bandwidth), in which the front left and right channels retain positional compatibility with two channel stereo. Furthermore, the discrete channel delivery possibilities of digital transmission and storage formats deal with the former problems of matrix encoding. If necessary, a truly separate two channel mix can be carried along with a surround mix of the same material. As will be explained later, this ITU standard does not define anything about the way that sound signals are represented or coded for surround sound, it simply states the layout of the loudspeakers. Most other things are open. So there is no 'correct' method of sound field representation or spatial encoding for this standard, but it is important to know that it was intended for the three front channels to be used for primary signal sources having clear directional attributes, whereas the rear/side channels were only ever intended as supporting ambience/effects/room channels to enhance the spatial effect. Many people are under the false impression that all-round localisation of sound images should be possible with such a layout, but in fact this is fraught with difficulty as discussed later. While it is not impossible to create side images, for example, they are nothing like as stable as front images, and the wide angle of the rear loudspeakers makes rear imaging problematic (tending to jump rapidly from one loudspeaker to the other as a signal is panned across, either using amplitude or time delay).

It appears likely that while movies are the driving force behind the current surround sound revolution, the music industry will benefit by riding on the band-wagon of the moving picture industry. Once systems are installed in people's homes, they will also be pleased to be able to play music releases and television broadcasts over the same system, perhaps without an accompanying picture. Although 'purist' sound engineers find it hard to accept that they must use a layout intended for movie reproduction, as it is not ideal for a variety of reasons to be discussed, most pragmatists realise that they are unlikely to succeed in getting a separate approach adopted for audio-only purposes and that they are best advised to compromise on what appears to be the best chance for a generation of enhancing the spatial listening experience for a large number of people.

1.5 Applications of spatial audio

A number of potential applications of spatial audio have been implied or stated above. In Table 1.1 applications are categorised in terms of the primary aim or purpose of spatial sound reproduction – in other words whether the application requires accurate rendering of sound sources and reflections over a 360° sphere, or whether the aim is primarily to deal in artistic/creative

Table 1.1 Categorisation of spatial audio applications

Application	Creative/artistic illusion	Accurate 3D rendering
Classical/live/natural music	•	(•)
Pop music	•	
Radio/TV sound (drama, documentaries, etc.)	•	(•)
Cinema sound	•	
Virtual reality	(•)	•
Computer games	•	(•)
Simulators/control systems (e.g. aircraft)		•
Conferencing/communication systems		•

illusion. Sometimes the distinction is not completely clear, and in such cases both have been shown.

Some of these categorisations could easily be disputed by others, but it was argued earlier that the primary aim of most commercial media production is not true spatial fidelity to some notional original sound field, although one might wish to create cues that are consistent with those experienced in natural environments. It is possible that the ability to render sources and reflections accurately in 360° would be considered valuable by recording engineers, but this creative freedom could be very hard to manage unless sophisticated and intelligent 'assistants' (as suggested recently by Andy Moorer in a keynote lecture to the Audio Engineering Society (Moorer, 2000)) were available to manage the translation from concept to implementation. Many sound engineers find it hard enough at the moment to know what to do with five channels.

In those applications categorised as being either mostly or completely concerned with accurate sound field rendering, the primary purpose of the system is related to some sort of spatial interaction with or human orientation in a 'reproduced world' or virtual environment. In such situations it is necessary to have quite precise auditory cues concerning the locations and movements of sound sources, so as to make sense of the scene presented and maybe interact with it. In some such applications it is important that the scene presentation adapts to the listener's movements, so that exploration of the scene is made possible in conjunction with visual cues. Such applications tend to lend themselves to an implementation of binaural technology, as this provides very precise control over the auditory cues that are created and the user is usually in a known physical relationship to the sound system, making it possible to control the signals presented to him or her accurately.

In the applications classed as primarily creative/artistic there may not be quite the same requirement for accuracy in spatial representation and sound field rendering, yet various spatial attributes of source material may need to be manipulated in a meaningful and predictable fashion. Listeners may not be located in a predictable physical relationship to the sound system making it more difficult to control the signals they are presented with. Such applications lend themselves more readily to multichannel audio systems, using a number of loudspeakers around the listening area. Clearly there are numerous 'crossover areas' between these two broad categories, where combinations of the two basic approaches might be adopted.

This book is somewhat more concerned with the latter of these two paradigms (multichannel audio systems), although it gives more than passing attention to the former. For a coverage more biased towards the former the reader is directed to the book *3D Sound for Virtual Reality and Multimedia* (Begault, 1994).

References

Begault, D. (1994). *3D Sound for Virtual Reality and Multimedia*. Academic Press.

Blauert, J. (1997). *Spatial Hearing*. MIT Press.

Blumlein, A. (1931). *British Patent Specification 394325*. Improvements in and relating to sound transmission, sound recording and sound reproducing systems.

Clark, H., Dutton, G. and Vanderlyn, P. (1958). The 'stereosonic' recording and reproducing system: a two-channel system for domestic tape records. *J. Audio Eng. Soc.*, **6**, 2, pp. 102–117.

Hertz, B. (1981). 100 years with stereo: the beginning. *J. Audio Eng. Soc.*, **29**, 5, pp. 368–372.

Moorer, J.A. (2000). Audio in the new millennium. *J. Audio Eng. Soc.*, **48**, 5, pp. 490–498.

North, A. and Hargreaves, D. (1997). Experimental aesthetics and everyday music listening. In *The Social Psychology of Music*, eds. North, A. and Hargreaves, D., Oxford University Press.

Steinberg, J. and Snow, W. (1934). Auditory perspectives – physical factors. In *Stereophonic Techniques*, pp. 3–7. Audio Engineering Society.

2 Spatial audio psychoacoustics

This chapter is concerned with the perception and cognition of spatial sound as it relates to sound recording and reproduction. It is not intended as an exhaustive review of spatial perception, as this has been very thoroughly done in other places, notably Blauert (1997) in his book *Spatial Hearing* and Moore (1989) in *An Introduction to the Psychology of Hearing*. Specifically, this chapter summaries those psychoacoustic phenomena that appear most relevant to the design and implementation of audio systems.

2.1 Sound source localisation

Most research into the mechanisms underlying directional sound perception conclude that there are two primary mechanisms at work, the importance of each depending on the nature of the sound signal and the conflicting environmental cues that may accompany discrete sources. These broad mechanisms involve the detection of timing or phase differences between the ears, and of amplitude or spectral differences between the ears. The majority of spatial perception is dependent on the listener having two ears, although certain monaural cues have been shown to exist – in other words it is mainly the *differences* in signals received by the two ears that matter.

2.1.1 Time cues

A sound source located off the 0° (centre front) axis will give rise to a time difference between the signals arriving at the ears of

Figure 2.1 The interaural time difference (ITD) for a listener depends on the angle of incidence of the source, as this affects the additional distance that the sound wave has to travel to the more distant ear. In this model the ITD is given by $r(\theta + \sin\theta)/c$ (where c = 340 m/s, the speed of sound, and θ is in radians).

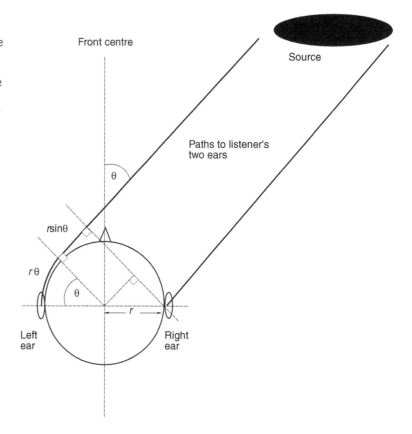

the listener that is related to its angle of incidence, as shown in Figure 2.1. This rises to a maximum for sources at the side of the head, and enables the brain to localise sources in the direction of the earlier ear. The maximum time delay between the ears is of the order of 650 µs or 0.65 ms and is called the binaural delay. It is apparent that humans are capable of resolving direction down to a resolution of a few degrees by this method. There is no obvious way of distinguishing between front and rear sources or of detecting elevation by this method, but one way of resolving this confusion is by taking into account the effect of head movements. Front and rear sources at the same angle of offset from centre to one side, for example, will result in opposite changes in time of arrival for a given direction of head turning.

Time difference cues are particularly registered at the starts and ends of sounds (onsets and offsets) and seem to be primarily based on the low frequency content of the sound signal. They are useful for monitoring the differences in onset and offset of the overall envelope of sound signals at higher frequencies. It is important to distinguish between the binaural delay resulting from a single source and the delay measured at each ear between two or more similar sound sources in different locations. The latter is a form of *precedence effect* and normally causes the brain to localise the sound towards the earlier of the two sources. This is discussed in more detail below.

Timing differences can be expressed as phase differences when considering sinusoidal signals. The ear is sensitive to interaural phase differences only at low frequencies and the sensitivity to phase begins to deteriorate above about 1 kHz. At low frequencies the hair cells in the inner ear fire regularly at specific points in the phase of the sound cycle, but at high frequencies this pattern becomes more random and not locked to any repeatable point in the cycle. Sound sources in the lateral plane give rise to phase differences between the ears that depend on their angle of offset from the 0° axis (centre front). Because the distance between the ears is constant, the phase difference will depend on the frequency and location of the source. (Some sources also show a small difference in the time delay between the ears at LF and HF.) Such a phase difference model of directional perception is only really relevant for continuous sine waves auditioned in anechoic environments, which are rarely heard except in laboratories. It also gives ambiguous information above about 700 Hz where the distance between the ears is equal to half a wavelength of the sound, because it is impossible to tell which ear is lagging and which is leading. Also there arise frequencies where the phase difference is zero. Phase differences can also be confusing in reflective environments where room modes and other effects of reflections may modify the phase cues present at the ears.

2.1.2 Amplitude and spectral cues

The head's size makes it an appreciable barrier to sound at high frequencies but not at low frequencies. Furthermore, the unusual shape of the pinna (the visible part of the outer ear) gives rise to reflections and resonances that change the spectrum of the sound at the eardrum depending on the angle of incidence of a sound wave. Reflections off the shoulders and body also modify the spectrum to some extent. A final amplitude cue that may be relevant for spherical wave sources close to the head is the level difference due to the extra distance travelled between the ears by off centre sources. For sources at most normal distances from the head this level difference is minimal, because the extra distance travelled is negligible compared with that already travelled.

The sum of all of these effects is a unique *head-related transfer function* or HRTF for every source position and angle of incidence, including different elevations and front–back positions. Some examples of HRTFs at different angles are shown in Figure 2.2. It will be seen that there are numerous spectral peaks and dips, particularly at high frequencies, and common features have been found that characterise certain

Figure 2.2 Monaural transfer functions of the left ear for several directions in the horizontal plane, relative to sound incident from the front; anechoic chamber, 2 m loudspeaker distance, impulse technique, 25 subjects, complex averaging (Blauert, 1997). (a) Level difference; (b) time difference. Courtesy of MIT Press.

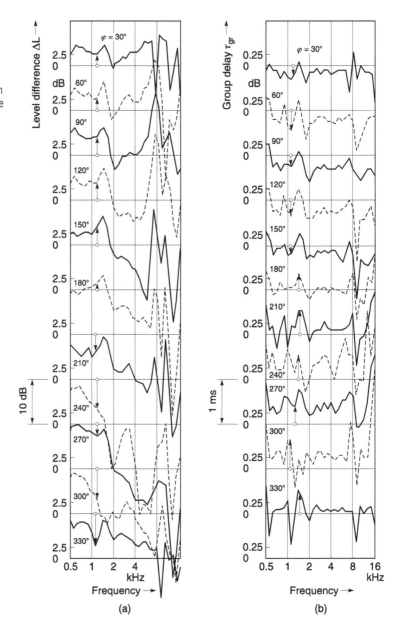

source positions. This, therefore, is a unique form of directional encoding that the brain can learn. Typically, sources to the rear give rise to a reduced high frequency response compared to those at the front, owing to the slightly forward facing shape of the pinna. Blauert has found evidence of so-called 'directional bands' which are regions of the frequency spectrum that appear boosted or attenuated for particular source positions in the median plane. A region around 8 kHz appears to correspond quite closely to overhead perception, whereas regions from 300–600 and 3000–6000 Hz seem to relate quite closely to frontal perception. Regions centred on about 1200 Hz and 12000 Hz appear to be closely related to rear perception. Besides these broad directional bands others have identified narrow peaks and notches corresponding to certain locations. Hebrank and Wright (1974), for example, identified the following broad relationships between spectral features and median plane localisation: front – a one octave notch with the lower end between 4 and 8 kHz and increased energy above 13 kHz; above – a one quarter octave peak between 7 and 9 kHz; behind – a small peak between 10 and 12 kHz with a decrease of energy above and below; front elevation varied with the lower cutoff frequency of a one octave notch between 5 and 11 kHz.

These HRTFs are superimposed on the natural spectra of the sources themselves. It is therefore hard to understand how the brain might use the monaural spectral characteristics of sounds to determine their positions as it would be difficult to separate the timbral characteristics of sources from those added by the HRTF. There is some evidence, though, that monaural cues provide some directional information and that the brain is capable of comparing monaural HRTFs with stored patterns to determine source location (Plenge, 1972, cited in Moore), or that Blauert's broad directional bands are evaluated by the brain to determine the regions of the spectrum in which the most power is concentrated (this is only plausible with certain sources, having reasonably flat or uniform spectra). Monaural cues are likely to be more detectable with moving sources, because moving sources allow the brain to track changes in the spectral characteristics that should be independent of a source's own spectrum. For lateralisation it is most likely to be *differences* in HRTFs between the ears that help the brain to localise sources, in conjunction with the associated interaural time delay. Monaural cues may be more relevant for localisation in the median plane where there are minimal differences between the ears.

A study of a few human pinnae will quickly show that, rather like fingerprints, they are not identical. They vary quite widely in shape and size. Consequently, so do HRTFs, which makes it difficult to generalise the spectral characteristics across large numbers of individuals. People that have tried experiments where they are given another person's HRTF, by blocking their own pinnae and feeding signals directly to the ear canal, have found that their localising ability is markedly reduced. After a short time, though, they appear to adapt to the new information. This has implications for binaural audio signal processing. Considerable effort has taken place, particularly over the last twenty years, to characterise human HRTFs and to find what features are most important for directional perception. If certain details of HRTFs can be simplified or generalised then it makes them much easier to simulate in audio systems, and for the results to work reasonably well for different listeners. There is some evidence that generalisation is possible, but people localise best with their own HRTFs. There are even known to be 'good localisers' and 'poor localisers', and the HRTFs of good localisers are sometimes found to be more useful for general application.

The so-called *concha* resonance (that created by the main cavity in the centre of the pinna) is believed to be responsible for creating a sense of externalisation – in other words a sense that the sound emanates from outside the head rather than within. Sound reproducing systems that disturb or distort this resonance, such as certain headphone types, tend to create in-the-head localisation as a result. This has led some researchers to attempt the design of headphones that stimulate the concha resonance from the front direction, outside the entrance of the ear canal, so as to superimpose certain individual listener hearing features on reproduced binaural signals (see Chapter 3), thereby improving frontal 'out-of-head' perception with headphone listening (Tan and Gan, 2000).

2.1.3 Binaural delay and various forms of precedence effect

As mentioned above, there is a distinct difference between the spatial perception that arises when two ears detect a single wavefront (i.e. from a single source) and that which arises when two arrivals of a similar sound come from different directions and are detected by both ears (as shown in Figure 2.3). The former gives rise to spatial perceptions based primarily on what is known as the 'binaural delay' (essentially the time-of-arrival difference that arises between the ears for the particular angle of incidence), and the latter gives rise to spatial perceptions based

Figure 2.3 Two instances of spatial perception. (a) A single source emitting a wavefront that is perceived separately by the two ears. Time-based localisation primarily determined by the binaural delay. Most relevant to headphone reproduction and natural listening. (b) Two sources in different locations emitting essentially the same signal, creating two wavefronts both of which are perceived by both ears (each wavefront separately giving rise to the relevant binaural delay). Time-based localisation primarily determined by the precedence effect or 'law of the first wavefront', which depends upon the relative delay and amplitude of the two signals. Most relevant to loudspeaker reproduction.

(a)

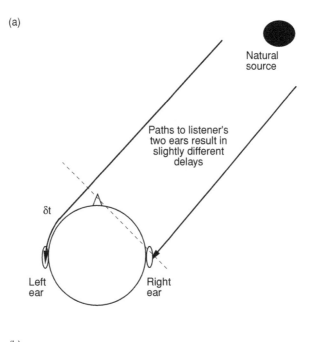

(b)

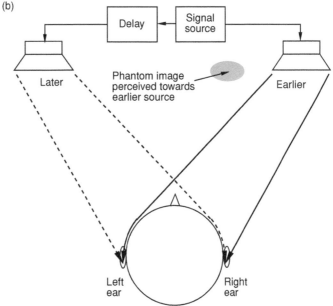

primarily on various forms of 'precedence effect' (or 'law of the first wavefront'). In terms of sound reproduction, the former may be encountered in the headphone presentation context where sound source positions may be implied by using delays between the ear signals within the interaural delay of about 0.65 ms. Headphones enable the two ears to be stimulated independently of each other.

In loudspeaker listening the precedence effect is more relevant, as a rule. The precedence effect is primarily a feature of transient sounds rather than continuous sounds. In this case there are usually at least two sound sources in different places, emitting different versions of the same sound, perhaps with a time or amplitude offset to provide directional information. Both ears hear all loudspeakers and the brain tends to localise based on the interaural delay arising from the earliest arriving wavefront, the source appearing to come from a direction towards that of the earliest arriving signal (within limits). This effect operates over delays between the sources that are somewhat greater than the interaural delay, of the order of a few milliseconds. Similar sounds arriving within up to 50 ms of each other tend to be perceptually fused together, such that one is not perceived as an echo of the other. The time delay over which this fusing effect obtains depends on the source, with clicks tending to separate before complex sounds like music or speech. The timbre and spatial qualities of this 'fused sound', though, may be affected.

One form of precedence effect is sometimes referred to as the Haas effect after the Dutch scientist who conducted some of the original experiments. It was originally identified in experiments designed to determine what would happen to the perception of speech in the presence of a single echo. Haas determined that the delayed 'echo' could be made substantially louder than the earlier sound before it was perceived to be equally loud, as shown in the approximation in Figure 2.4. The effect depends considerably on the spatial separation of the two or more sources

Figure 2.4 A crude approximation of the so-called 'Haas effect' showing the relative level required of a delayed reflection (secondary source) for it to appear equally loud to an earlier primary source.

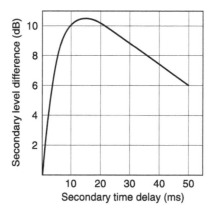

involved. This has important implications for recording techniques where time and intensity differences between channels are used either separately or combined to create spatial cues. A useful review of the precedence effect in sound localisation was given by Wallach, Newman and Rosenzweig (1949).

2.1.4 Time–intensity-based localisation related to sound reproduction

Time and intensity differences between the ears can be traded against each other for similar perceived directional effect. A number of researchers have investigated this issue and come up with different values for the number of dB level differences that can be used to compensate for a certain number of microseconds timing difference in the binaural delay. This seems to depend on the nature of the source stimulus to some extent. A summary of this effect, after Madsen, is shown in Figure 2.5. Here the time–intensity trade-off ceases to work once the delay reaches the maximum binaural delay, leaving the perceived direction always towards the earlier sound until the two sounds are perceived separately.

Whitworth and Jeffress (1961) attempted to measure the trade-off between time and intensity difference in source localisation, using binaural headphones (thus presenting each ear with an

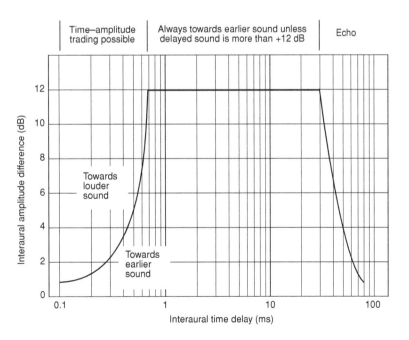

Figure 2.5 Binaural time–intensity trading effects (after Howard and Angus, originally from Madsen).

independent signal). They commented on the enormous disparity between the trade-off values found by different workers, ranging between the extremes of 120 µs per dB and 1.7 µs per dB. The type of source used in each of these two examples was quite different – the first used low-level clicks and the second used mainly low-frequency tones at a higher level. The implication is that a different mechanism may be used for locating tones from that which is used for transients – with tones appearing to exhibit a much smaller trading ratio than transients. Jeffress further hypothesised that there were two images: the 'time image' and the 'level image', which were not directly interchangeable.

In fact the situation is more complicated than this, since Harris (1960) was able to show that clicks filtered so as to contain only energy below 1500 Hz resulted in a time–intensity trade-off of 25 µs/dB, whereas the same clicks highpass filtered so as to remove the LF energy resulted in about 90 µs/dB. There is also a distinct suggestion from some workers that it may not be possible to trade time and intensity differences with complete freedom, and this has important relevance to stereo microphone systems. Hafter and Carrier (1972), for example, were able to show that listeners could tell the difference between signals which had been centred in the image by offsetting a time difference with a level difference, and a signal which was naturally centred because no time or level difference had been introduced. Moore (1989) has also concluded that time and level differences are not truly equivalent, but notes that this may only be a phenomenon important in the laboratory, when using carefully selected signals, since in real life it is clear that both time and level difference work in conjunction to produce a single clearly-defined image.

Yost *et al.* (1971) showed how the LF content of transients was considerably more important for localisation than the HF content. It was suggested that because LF sounds excite the end of the basilar membrane furthest from the oval window within the inner ear, and because the damping of the membrane is somewhat lower at this end, the decay of the stimulus would last for longer and thus provide a more prolonged flow of information to the brain. (In a number of reports it has been shown that the result of stimulating the inner ear with a transient is actually a decaying sinusoidal waveform on the basilar membrane, and thus the brain really compares the relative phases of these waveforms as the transient decays, rather than simply the arrival times of the edge of the transient, in order to determine the position of the source.) Yost's paper begins with the statement that in a realistic acoustic environment, steady-state sounds do not provide reliable information about the location of a sound source – that it is transient information which offers the more relevant directional cue.

This all rather supports Jeffress' hypothesis that there are two mechanisms for localisation – one being governed largely by the low-frequency content of transient information (below about 1500 Hz) and relying solely on time difference, whereas another mechanism is more dependent on the level of the signal, working on both time and level differences across the whole frequency range.

A similar form of trading can take place between multiple sources, according to some sort of precedence effect. For example, Williams (see Chapters 6 and 7) bases his family of near-coincident microphone arrays on some time–intensity trading curves derived from experiments on stereo loudspeaker signals conducted at the Technical University of Lyngby, Denmark (Simonsen, 1984), as shown in Figure 2.6. Here we see curves relating the necessary combinations of time and level difference to make a panned mono source appear to be at either 10, 20 or 30 degrees off centre in a two loudspeaker stereo system (30 degrees would be fully left or right). We see that in this experiment a time difference of about 1.1 ms or a level difference of about 15 dB (or some in-between combination of the two) is required to make a source appear to be fully to one side.

In experiments using a five loudspeaker configuration according to the ITU-R BS 775 standard (the standard five-channel

Figure 2.6 Time and level difference combinations between two front loudspeakers at ±30° in a typical listening room, related to perceived location of phantom image (after Williams). The small circles represent the data points determined by Simonsen (1984), whereas the curves were interpolated by Williams. Signals were speech and maracas.

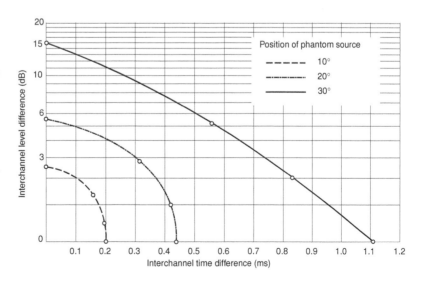

surround configuration), Geoff Martin and his colleagues from McGill University found that the time delays and amplitude differences required to make sources appear to be in certain places depended on which loudspeaker pair of the five was involved (Martin *et al.*, 1999). Between the rear loudspeaker pair a time delay of only about 0.6 ms was required for a signal to appear fully to one side – roughly half that required for the front channels (where results were similar to Simonsen). This is likely to be due to the wider angle subtended by the rear loudspeakers in such a configuration (about 120° as opposed to 60°). For signals delayed between the front-left and rear-left loudspeakers the results were unconvincing owing to the difficulty of localising signals on the basis of time delays between two sources at the same side of the head. Their conclusion was that amplitude differences between channels provided more stable images than time differences.

Figure 2.7 Image location and stability between pairs of loudspeakers in a four channel square array, using amplitude panning (Ratliffe, 1974, courtesy of BBC Research and Development).

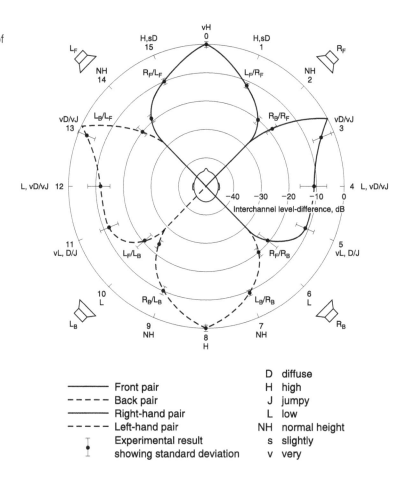

——— Front pair	D diffuse
– – – Back pair	H high
——— Right-hand pair	J jumpy
– – – Left-hand pair	L low
Experimental result	NH normal height
showing standard deviation	s slightly
	v very

Paul Ratliffe at the BBC Research Department conducted similar experiments in the 1970s, but in this case for a square arrangement of loudspeakers intended for quadraphonic reproduction. Figure 2.7 shows his primary conclusions (Ratliffe, 1974). He concluded that phantom images based on amplitude differences between side pairs were poorly localised and that sources appeared to jump rapidly from front to back rather than panning smoothly down the sides. This is attributed to the difficulty of creating interaural differences of any significance from differences between loudspeaker pairs to the same sides of the head.

2.1.5 Effects of reflections

This topic is expanded to some extent in Chapter 5, when considering listening environments, but it is worth noting here the general principle that the level and amplitude of reflections arising from sources in listening spaces also affects spatial perception significantly. This is a huge subject that could fill another book on its own, but it is mentioned in various relevant places in this book. Reflections in the early time period after direct sound (up to 50–80 ms) typically have the effect of broadening or deepening the spatial attributes of a source. They are unlikely to be individually localisable but do affect spatial perception. In the period up to about 20 ms they can cause severe timbral coloration if they are at high levels. After 80 ms they tend to contribute more to the sense of envelopment or spaciousness of the environment (see below). David Griesinger has stated a number of times that reflections between 50–150 ms are problematic in sound reproduction, serving mainly to detract from intelligibility and clarity. Although they exist in real spaces he is inclined to minimise them in artificial reverberation devices for use in sound recording, so he typically concentrates on the period up to 50 ms to simulate depth and source broadening and after 150 ms to create spaciousness.

2.1.6 Interaction between hearing and other senses

Some spatial cues are context dependent and may be strongly influenced by the information presented by other senses, particularly vision. Learned experience leads the brain to expect certain cues to imply certain spatial conditions, and if this is contradicted then confusion may arise. For example, it is unusual to experience the sound of a plane flying along beneath one, but the situation can occasionally arise when climbing mountains. Generally one expects planes to fly above, and most people will look up or duck when played loud binaural recordings of planes flying over, even if the spectral cues do not imply this direction.

It is normal to rely quite heavily on the visual sense for information about events within the visible field, and it is interesting to note that most people, when played binaural recordings of sound scenes without accompanying visual information or any form of head tracking, localise the scene primarily behind them rather than in front. In fact obtaining front images from any binaural system using headphones is surprisingly difficult. This may be because one is used to using the hearing sense to localise things where they cannot be seen, and that if something cannot be seen it is likely to be behind. In the absence of the ability to move the head to resolve front–back conflicts the brain tends to assume a rear sound image. So-called 'reversals' in binaural audio systems are consequently very common.

The congruence expected between audible and visual scenes in terms of localisation of objects seems to depend on the level of experience of the subject. In experiments designed to determine the degree of directional distortion acceptable in sound/picture systems it was found that a 11° mismatch was annoying for experts but this was loosened to 20° for naïve subjects (Komiyama, 1989).

Begault provides an interesting review of a number of these issues in his paper 'Auditory and non-auditory factors that potentially influence virtual acoustic imagery' (Begault, 1999).

2.1.7 Resolving conflicting cues

In environments where different cues conflict in respect of the implied location of sound sources, the hearing process appears to operate on a sort of majority decision logic basis. In other words it evaluates the available information and votes on the most likely situation, based on what it can determine. Auditory perception has been likened to a hypothesis generation and testing process, whereby likely scenarios are constructed from the available information and tested against subsequent experience (often over a very short time interval). Context dependent cues and those from other senses are quite important here. Since there is a strong precedence effect favouring the first arriving wavefront, the direct sound in a reflective environment (which arrives at the listener first) will tend to affect localisation most, while subsequent reflections may be considered less important. Head movements will also help to resolve some conflicts, as will visual cues. Reflections from the nearest surfaces, though, particularly the floor, can aid the localising process in a subtle way. Moving sources also tend to provide more information than stationary ones, allowing the brain to measure changes in the received information that may resolve some uncertainties.

Also important in this respect appears to be the issue of auditory scene formation and object recognition, discussed below. Cues that do not fit the current cognitive framework may be regarded as more suspect than those that do.

2.2 Distance and depth perception

Apart from lateralisation of sound sources, the ability to perceive distance and depth of sound images is crucial to our subjective appreciation of sound quality. Distance is a term specifically related to how far away an individual source appears to be, whereas depth can describe the overall front-back distance of a scene and the sense of perspective created. Individual sources may also appear to have depth.

A number of factors appear to contribute to distance perception, depending on whether one is working in reflective or 'dead' environments. Considering for a moment the simple differences between a sound source close to a listener and the same source further away, the one further away will have the following differences:

1   Quieter (extra distance travelled).
2   Less high frequency content (air absorbtion).
3   More reverberant (in reflective environment).
4   Less difference between time of direct sound and first floor reflection.
5   Attenuated ground reflection.

Numerous studies have shown that absolute distance perception, using the auditory sense alone, is very unreliable in non-reflective environments, although it is possible for listeners to be reasonably accurate in judging relative distances (since there is then a reference point with known distance against which other sources can be compared). In reflective environments, on the other hand, there is substantial additional information available to the brain. The ratio of direct to reverberant sound is directly related to source distance, the reverberation time and the early reflection timing tells the brain a lot about the size of the space and the distance to the surfaces, thereby giving it boundaries beyond which sources could not reasonably be expected to lie.

Very close to the listener there are also some considerable changes in the HRTF spectra of sources, as reviewed by Huopaniemi (1999). There appears to be a degree of low frequency and high frequency increase in the interaural level difference between the ears for sources at very close distances.

2.3 Apparent source width

The subjective phenomenon of apparent or auditory source width (ASW) has been studied for a number of years, particularly by psychoacousticians interested in the acoustics of concert halls. (For a useful review of this topic, see Beranek (1997): *Concert and Opera Halls: How They Sound*). ASW relates to the issue of how large a space a source appears to occupy from a sonic point of view (ignoring vision for the moment), as shown in Figure 2.8, and is best described as a 'source spaciousness' phenomenon. Early reflected energy in a space (up to about 80 ms) appears to modify the ASW of a source by broadening it somewhat, depending on the magnitude and time delay of early reflections. Concert hall experiments seem to show that subjects

Figure 2.8 Graphical representation of the concept of apparent source width (ASW). (a) Small ASW. (b) Larger ASW.

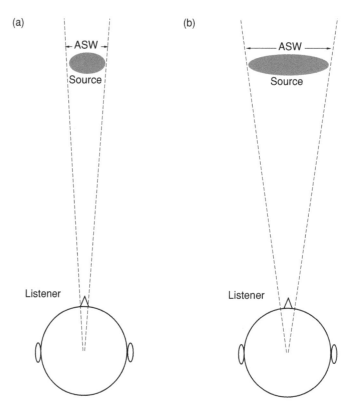

prefer larger amounts of ASW, but it is not clear what is the *optimum* degree of ASW (presumably sources that appeared excessively large would be difficult to localise and unnatural).

ASW has been found to relate quite closely to a binaural measurement known as interaural cross correlation, which (put crudely) measures the degree of similarity between the signals at the two ears. There are various ways of measuring and calculating IACC, described by different authors, and it can be measured in a number of different frequency bands and time windows. Early IACC, that is measured over a time window up to about 80 ms after the direct sound, appears to correlate best to ASW in concert halls.

This subjective attribute of source width arises also in sound reproduction and is possibly associated with image blur. Some experts, such as Griesinger (1999), are of the opinion that ASW is of little relevance in small room acoustics. It is often hard to say whether a reproduced sound image is wide or just rather diffuse and difficult to localise. Furthermore, individual source width should be distinguished from overall 'sound stage width' (in other words, the distance perceived between the left and right limits of the stereophonic scene). It is unclear whether the same preference for larger ASW exists in reproduced sound as in concert hall acoustics. Anecdotal experience suggests that precise 'tight' imaging capabilities of audio systems are quite important to pop mixing engineers, but at the same time it is common to use microphone techniques in classical music recording that have the effect of adding 'air' or 'space' around sources so that they do not appear to emanate from a single point. The issue is probably more one of aesthetics and convention in pop balancing.

2.4 Envelopment and spaciousness

The terms envelopment and spaciousness, and sometimes 'room impression', arise increasingly frequently these days when describing the spatial properties of sound reproducing systems. They are primarily related to environmental spatial impression, and are largely the result of reflected sound – almost certainly late reflected sound (particularly lateral reflections after about 80 ms). The problem with such phenomena is that they are hard to pin down in order that one can be clear that different people are in fact describing the same thing. It has been known for people also to describe envelopment and spaciousness in terms that relate more directly to sources than environments.

Spaciousness is used most often to describe the sense of open space or 'room' in which the subject is located, usually as a result of some sound sources such as musical instruments playing in that space. It is also related to the sense of 'externalisation' perceived – in other words whether the sound appears to be outside the head rather than constrained to a region close to or inside it. Envelopment is a similar term and is used to describe the sense of immersivity and involvement in a (reverberant) soundfield, with that sound appearing to come from all around. It is regarded as a positive quality that is experienced in good concert halls. Difficulties arise with defining these concepts in reproduced sound, particularly in the case of somewhat artificial balances that do not present a natural sound stage, because sources are sometimes placed in surround balances such that direct sounds appear to envelop the listener, rather than reverberant sound. The distinction is probably only of academic interest, but is important in attempting to evaluate the performance of systems and recording techniques, and in the search for physical measurements that relate to different subjective attributes.

Researchers such as Bradley have attempted to define measures that relate well to listener envelopment (LEV) in concert halls (Bradley and Souloudre, 1995). Measures such as 'lateral fraction' (LF) and 'lateral gain' (LG_{80}) have been proposed, relating to the proportion of late lateral energy in halls compared with an omnidirectional measurement of sound pressure at the listening position. It is not yet known to what degree these are relevant in small room reproduction, but Griesinger (1999) provides a useful overview of his work on these issues. He has proposed that decorrelation between the signals fed to multiple replay loudspeakers is important at low frequencies if spaciousness is to be perceived (below 700 Hz), but that greater correlation may be needed at higher frequencies. The interaction between loudspeakers and listening rooms at low frequencies is also an important variable in this respect (see Chapter 5).

Mason, as well as Griesinger, has attempted to develop an objective measurement that relates closely to subjective spatial impression (Mason and Rumsey 2000). In both studies it appears that conventional IACC measurements are inadequate predictors of spatial qualities for reproduced sound in small rooms, particularly at low frequencies. More important, it seems, is the *fluctuation* in interaural time delay (or IACC) that results from the interaction between multiple sources and their reflections. The importance of different rates and magnitudes of these fluctuations for different forms of spatial impression is currently under investigation.

2.5 Naturalness

While perhaps rather too general a term to be measurable directly by any physical means, the subjective attribute of 'naturalness' appears to arise over and over again in subjective data relating to spatial sound reproduction. It appears to be relatively independent of other factors and relates to the subject's perception of the degree of 'realism' or 'trueness to nature' of the spatial experience (Berg and Rumsey, 2000a). In some experiments it seems to be by far the most important factor in determining overall preference in sound quality (e.g. Mason and Rumsey, 2000). Possibly it is mainly an evaluative or emotive judgement, and it may consist of an optimum combination of other sub-factors, and it may have a strong timbral component and be highly context dependent. It is mentioned here simply because it is important to recognise that attention to the spatial factors described earlier (e.g. ASW, LEV, locatedness) is only part of what may make reproduced sound appear natural or artificial.

The majority of spatial cues received in reproduced sound environments are similar to those received in natural environments, although their magnitudes and natures may be modified somewhat. There are, nonetheless, occasional phenomena that might be considered as specifically associated with reproduced sound, being rarely or never encountered in natural environments. The one that springs most readily to mind is the 'out of phase' phenomenon, in which two sound sources such as loudspeakers or headphones are oscillating exactly 180° out of phase with each other – usually the result of a polarity inversion somewhere in the signal chain. This creates an uncomfortable sensation with a strong but rather unnatural sense of spaciousness, and makes phantom sources hard to localise. The out-of-phase sensation never arises in natural listening and many people find it quite disorientating and uncomfortable. Its unfamiliarity makes it hard to identify for naïve listeners, whereas for expert audio engineers its sound is unmistakeable. Naïve listeners may even quite like the effect, and extreme phase effects have sometimes been used in low-end audio products to create a sense of extra stereo width.

Audio engineers also often refer to problems with spatial reproduction as being 'phasy' in quality. Usually this is a negative term that can imply abnormal phase differences between the channels, or an unnatural degree of phase difference that may be changing with time. Anomalies in signal processing or microphone technique can create such effects and they are unique to reproduced sound, so there is in effect no natural anchor or reference point against which to compare these experiences.

2.6 Some subjective experiments involving spatial attributes of reproduced sound

2.6.1 *Subjective assessment of multichannel reproduction*

One of the few examples of spatial subjective quality tests carried out during the previous intense period of interest in multichannel surround reproduction is the work of Nakayama *et al.* (1971). He studied the subjective effects of 1–8 channel reproductions in an anechoic chamber using recordings made in a concert hall with unidirectional microphones in the same arrangement as the reproducing loudspeakers. Other microphone arrangements such as an MS pair and a close multimicrophone balance were also used (see Chapter 6). The microphone array was used at three different distances from the orchestra.

Two different approaches were used in the subsequent subjective assessment, in which 13 different speaker arrangements ranging from 1 to 8 channels were presented. In a single-stimulus experiment (one sound example graded at a time) listeners made a preference judgement on a seven-point scale, ranging from 'very good' to 'very bad'. In a paired-stimulus experiment (involving the comparison of pairs of sound examples) listeners were asked to judge the similarity between stimuli, also on a seven-point scale ranging from 'just the same' to 'quite different'. A distance scale for preference was constructed from the quality judgements and the similarity data were converted to similarity distances between all combinations and subjected to multidimensional analysis (MDA).

The subjective factors they identify as important in explaining the results are interpreted as (a) 'depth of image sources', (b) 'fullness', (c) 'clearness'. An examination of their results suggests that 'fullness' is very similar to what others have called 'envelopment', as it is heavily loaded for reproductions involving more loudspeakers to the sides and rear of the listener, and weak for two-channel frontal stereo. It appeared to be greatest in a four-channel reproduction when the side loudspeakers were located between about 50 and 60° off front centre (two front speakers at ±15°). 'Depth of sources' seems in fact to be more like 'nearness' or 'closeness' of sources when one reads the authors' comments, providing a good example of the difficulties of language and translation in comparing such results with others. It changed greatly as the recording position of the microphones was moved closer to the orchestra, as one might expect. 'Clearness' was found to relate closely to the measured concert hall acoustics parameter D50 (Definition or *Deutlichkeit*), and is thus clearly an indication of direct to reverberant ratio. D50, put simply, compares the sound energy arriving in the first 50 ms with later arriving energy.

They also formulated an equation that related the quality ratings of listeners to the three attributes by weighting the factors appropriately, based on a least-squares solution which fitted values from the three scales to the observed quality values. Their equation suggests that 'fullness' ('envelopment'?) was weighted most strongly in this equation, followed by 'depth of sources', followed by 'clearness', which is most revealing.

The authors' concluding remarks are worth noting with regard to the problem of assessing 'non-natural' recorded material.

> Needless to say, the present study is concerned with the multichannel reproduction of music played only in front of the listeners, and proves to be mainly concerned with extending the ambience effect... In other types of four-channel reproduction the localisations of image sources are not limited to the front. With regard to the subjective effects of these other types of reproduction, many further problems, those mainly belonging to the realm of art, are to be expected. The optimisation of these might require considerably more time to be spent in trial, analysis and study.

2.6.2 Perceived quality of sound reproducing systems

Some of the most well-known and in-depth investigations into perceived quality of sound reproducing systems were conducted by Gabrielsson and others. For example, Gabrielsson and Sjören (1979) conducted a range of experiments aiming, among other things, 'to find out and interpret the meaning of relevant dimensions entering into perceived sound quality'. They conducted tests on headphones, loudspeakers and hearing aids, in mono on loudspeakers and stereo on headphones. Subjects were asked (a) to rate stimuli on a large number of adjective scales that had previously been selected by a group of sound engineers from a longer list; (b) to rate the similarity between pairs of stimuli; (c) to provide free verbal descriptions of a sample of stimuli.

The adjective ratings were analysed using principal components analysis (PCA) in an attempt to isolate a limited number of quality 'factors'. PCA achieves this by looking for correlations between the multiple adjective ratings and then offering a limited number of principal factors or components which represent the main perceptual dimensions on which the adjectives seem to be most correlated. The factor weightings given to each adjective show how each 'scored' under each perceptual factor (they extracted three factors in the loudspeaker test and five in the headphone test), and this assists in interpreting the meaning of each factor. While the majority of adjective scales related to timbral and other attributes, a number related at least partially to the spatial attributes of reproduction. Terms such as 'distant/near', 'diffuse', 'closed/shut-up', 'airy', 'confined to a point', 'feeling of room', 'blurred', 'open', could all be considered spatial attributes, and scored high weightings on one of the factors which was interpreted by the authors as 'a general quality factor emphasizing clearness/distinctness, feeling of space and nearness in the reproduction'. In the headphone experiment one can isolate two factors from the five that may represent spatial attributes: the authors report that Factor II was interpreted as 'clearness/distinctness', and received high factor loadings for adjectives such as 'clear', 'pure/clean', 'true-to-nature' and 'feeling of presence', balanced up with strong negative factor loading for 'diffuse'. Factor V is characterised as 'feeling of space', showing a strong negative factor loading for the opposite 'closed/shut-up'. Factors II and V were also found to have a modest correlation (0.45) between them. In the hearing aid tests, the factor 'nearness' came out in one test.

The authors also looked for relationships between listeners' ratings of the two terms 'pleasant' and 'natural/true to nature' and the main factor loadings. In relation to the 'feeling of space' factor these terms appear loaded on the 'open/airy' side. With the 'nearness' factor the balance is in favour of 'near' rather than 'distant' (although not conclusively), and with the 'clearness/distinctness' factor the high loadings are towards the 'clear/distinct' side.

These experiments suggest strongly that spatial attributes are at least one of the main factors determining quality ratings in sound reproduction, and that there is a degree of consensus among listeners as to what spatial attributes are preferred.

2.7 Cognitive issues in sound space perception

Cognitive issues in sound listening concern the higher level interpretative aspects of the brain's function, and relate to the ways in which people make sense of their environment and experiences.

2.7.1 Grouping and streaming of sound cues in auditory scene analysis

Of increasing interest in psychoacoustics is the issue of so-called 'auditory scene analysis'. This term, also the title of a major book by Albert Bregman (Bregman, 1990), describes the process by which the brain groups sound stimuli and forms objects out of the basic features presented. The ability of the cognitive processes in the brain to create and label auditory 'objects' based on their constituent sonic components may help it to decide which objects belong together in a so-called 'scene' (the complex auditory image of a space) and how they relate to other elements in the scene. This enables the brain to build a model of the environment that the senses perceive. From the complex collection of frequency components and time-related features sent from the ears, the brain has to decipher meaningful information that can be used to determine, for example 'oboe slightly off centre, among other instruments in the woodwind section of an orchestra, towards the back of a somewhat reverberant hall'. The feat of signal processing and associative memory involved here is quite remarkable.

The means by which auditory concepts are formed seems to be related to perceptual streaming. Put crudely, certain time and frequency relationships between signal components, the details of which are too lengthy to study in detail here, tend to lead the brain to link them to form a common object. Where that object has been previously recognised and labelled, memory enables one to identify it. Higher levels of perception are often modelled in terms of pattern recognition, of which this streaming process is a part. Gestalt psychology has described many of the elements of perceptual organisation, as summarised succinctly by Moore (1989, pp. 245–253).

The consideration of objects and scenes leads one to consider the issue of differences between discrete objects and environmental cues in spatial audio. While considerable attention has been directed at the localisation of individual sources or objects in auditory perception, in fact much of the reflected sound in an enclosed space is not specifically localisable and tends to create a diffuse background to the scene. In any discussion of the spatial attributes of sound reproduction it may be helpful to group the sensations resulting from auditory stimuli into source attributes and environment attributes, and possibly also into attributes that describe a group of sources that are perceived as a single entity (such as the string section of an orchestra). The environment attributes are often described subjectively in terms of 'envelopment', 'spaciousness', 'room impression' and so on.

Source attributes tend to be described in terms of their location, apparent size and distance (Berg and Rumsey, 1999). Subjective descriptions such as 'depth' can be ambiguous, as they may relate to the depth of the scene as a whole or to an individual element or group of elements within it, requiring clarity in definition (Mason *et al.*, 2000).

David Griesinger has put forward theories relating to the way in which spatial features are grouped and perceived in reproduced sound environments, and he has presented hypotheses concerning the physical cues that control different forms of what he calls 'spatial impression' (Griesinger, 1997). He asserts that the spaciousness associated with a source image may be perceived as part of the source itself, linked directly to it, in which case it may be quite reasonable to describe a source rather than an environment as having some form of 'spaciousness'. There is an important link here between auditory streaming and what Griesinger has termed CSI (continuous spatial impression), ESI (early spatial impression) and BSI (background spatial impression). Griesinger asserts that when a direct sound source is continuous and cannot be split into separate events, the interaction with reflected energy and the interaural fluctuations in amplitude and time delay that result can give rise to a sense of full envelopment or spaciousness that appears to be connected to the sound (CSI). ESI, on the other hand, is related to separable sound events that form a foreground stream, where reflected energy arrives during the sound event and within 50 ms of its end. ESI, it is claimed, is not fully enveloping and is perceived as occupying much the same location as the source itself, contributing to image broadening. It is the spatial impression experienced in small rooms. In large spaces, or in other situations where much reflected energy arrives more than 50 ms after the ends of sound events, BSI results. Spatially diffuse, late reflected energy of this type results in good envelopment, but BSI is not bound to the source that created it (so it is in effect an environment cue). In assessing reproduced sound using typical listening rooms with short reverberation times, the BSI is almost certain to be provided by the recording rather than the room. BSI can probably be assessed subjectively using terms that relate more to environments, whereas CSI and ESI may require hybrid terms that relate to the spaciousness of sources.

2.7.2 A hierarchy of subjective attributes?

In our analysis of the subjective features of reproduced sound it may be helpful to create a hierarchical tree of attributes, in order to make the communication of meaning clear and to help in the

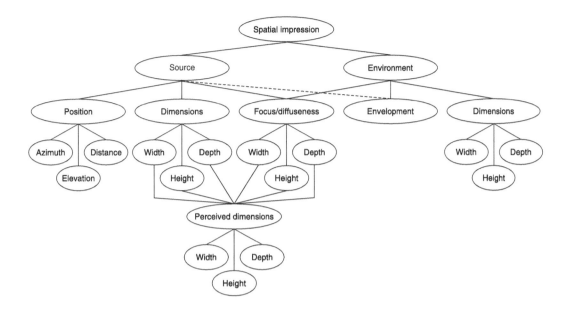

Figure 2.9 Proposed hierarchy of spatial attributes for use in subjective analysis (Mason, 1999).

study of relationships between physical attributes of sound systems and related subjective judgements. Mason has attempted such a structure (Mason, 1999), shown in Figure 2.9, as a basis for development.

2.7.3 Judgements versus sentiments

In studies of the subjective effects of spatial sound reproduction, it is sometimes useful to distinguish between what Nunally and Bernstein (1994) referred to as judgements and sentiments. Judgements, he suggested, were human responses or perceptions essentially free of personal opinion or emotional response and could be externally verified (such as the response to questions like 'how long is this piece of string?', or indeed 'what is the location of this sound source?'). Sentiments, on the other hand, could be said to be preference related or linked to some sort of emotive response and cannot be externally verified. Obvious examples are 'like/dislike' and 'good/bad' forms of judgement. Of course there are numerous examples one can think of that do not fit quite so neatly into either of these categories.

In spatial audio, provided we can define clearly enough what we mean by terms like envelopment (see below), and train subjects to appreciate their operational definition in terms of anchor stimuli or examples, we can probably reasonably well assume that their responses will be judgements. But in many cases it is hard to fulfil Nunally and Bernstein's criterion of 'externally verifiable' unless we have a physical measurement that relates closely to the subjective phenomenon. In any case, the issue of external verification might be questioned by some for the purpose of sound system evaluation as there seems little point in subjective assessment of a phenomenon if one can simply measure it directly, except perhaps to find out more about the incongruities between perception and the physical world.

With localisation the matter of external verification is somewhat easier, because with naturally occurring sources one can generally verify the location, enabling one to say how accurately a subject responded in each case. (The so-called 'auditory event' may appear to occur in a different location from that of the source itself.) The situation is complicated when one is synthesising the cues required to place a sound source in a reproduced environment (say by HRTF manipulation), as although one may know where one *intends* the source to be located there is no physical location to measure or verify.

In experiments designed to determine how subjects described spatial phenomena in reproduced sound systems, including surround sound, Berg and Rumsey (2000a, b) used a method known as repertory grid technique to elicit verbal scales from a number of subjects based on their perceptions of triads of different stimuli (representing different spatial reproduction modes). In order to separate descriptive attributes or constructs from emotional and evaluative ones (similar to Nunally and Bernstein's separation of judgements and sentiments) they used a form of verbal protocol analysis that filtered terms or phrases according to a classification method. Descriptive features could then be analysed separately from emotional responses, and relationships established between them in an attempt to determine what spatial features were most closely related to positive emotional responses. In this experiment it seemed that high levels of envelopment and room impression created by surround sound, rather than accurate imaging of sources, were the descriptive features most closely related to positive emotional responses.

2.8 The source–receiver signal chain

In sound recording and reproduction there are numerous components in the chain between original sound source and ultimate human receiver. Each of these can potentially contribute to the ultimate spatial effect, either negatively (because of some distortion or other) or positively (through some intentional optimisation of signal qualities). Some of these elements are depicted in Figure 2.10.

Figure 2.10
Recording/reproduction signal
chain showing elements that
could potentially affect the
spatial qualities of the sound,
either intentionally or
unintentionally.

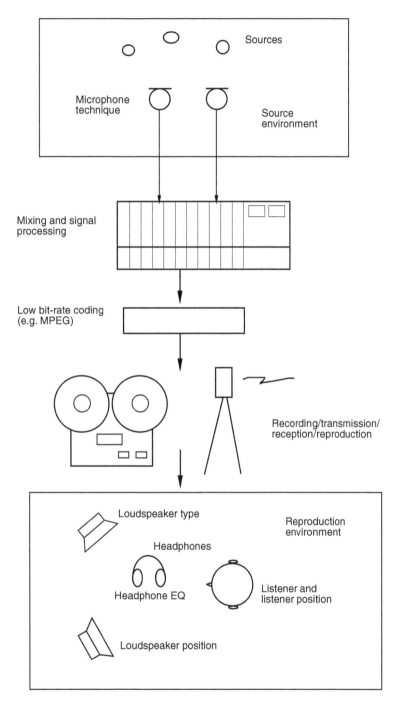

One has the source environment and its spatial characteristics to consider. In some recording circumstances, such as some close miked pop recording, it is not desirable to capture these to any significant degree since the spatial qualities of the balance will be largely determined by artificial panning and effects. In classical balancing, or other forms of recording where the original space is important, the microphone technique will often be designed to capture appropriate spatial cues from the source environment, and the remainder of the signal chain should be designed to convey these accurately. Inadequacies in the source environment and other commercial recording/production factors may encourage recording engineers to minimise the pickup of the natural spatial features of a venue, or to use microphone techniques that create a sense of spatiality that is different from that experienced by a listener in a 'good' seat. Artificial reverberation is often used in these cases to complement or even override that of the hall, and in such cases it is the spatial qualities of the artificial reverberation that will dominate the end result.

Mixing and signal processing equipment can be used to introduce artificial panning and spatial effects to multiple monophonic sources so as to 'spatialise' them. In some 3D audio systems this is done using binaural algorithms that simulate HRTF and room modelling characteristics, but the result is then suitable mainly for headphone listening or some form of transaural loudspeaker reproduction (see Chapter 3). In most music recording for mainstream release this spatialisation is currently performed using relatively crude amplitude panning techniques together with artificial effects, although there is increasing interest in more psychoacoustically sophisticated tools.

Distortions present in the signal chain between the recording and reproducing environment can contribute to major changes in the spatial characteristics of the signal. In the days of analogue recording and transmission this was a bigger problem than it is now, since phase and frequency response anomalies were commonplace. These could particularly affect stereo imaging, and were capable of destroying the subtle spectral and timing cues required for good binaural imaging. Digital signal chains are more reliable in this respect and such major distortions are less likely to be encountered in recording and transmission equipment. The major component of digital systems likely to cause problems is any low bit rate coding that may be used to reduce the data rate of the signal for storage or transmission. Some such systems use algorithms that attempt to reduce the spatial information content of multichannel signals by exploiting interchannel redundancy (similar information between channels), and using forms of 'joint stereo coding' that simplify the representation of spatial content. This is most likely to be problematic at very low bit rates.

The mode of reproduction (e.g. headphones or loudspeakers), the types of transducers used, their location and the influence of the listening room acoustics will all affect the perceived spatial quality of the reproduced signal. The listener's location with relation to the transducers is also a major factor, as indeed are the listeners themselves! This helps to show just how many factors can potentially influence the end result, and while some can be controlled or standardised many others cannot. One ought to allow for different listener positions, for example, and one's spatial masterpiece ought to work reasonably well in a variety of different rooms with different sorts of loudspeakers. It might therefore be reasonable to suggest that the most successful spatial audio technique or system is one that performs satisfactorily under a wide range of different end-user conditions and for a wide range of different listeners.

References

Begault, D. (1999). Auditory and non-auditory factors that potentially influence virtual acoustic imagery. In *Proceedings of the AES 16th International Conference, Rovaniemi, 10–12 April*, pp. 13–26. Audio Engineering Society.

Beranek, L. (1996). *Concert and Opera Halls: How They Sound*. Acoustical Society of America, Woodbury, NY.

Berg, J. and Rumsey, F. (1999). Spatial attribute identification and scaling by repertory grid technique and other methods. In *Proceedings of the AES 16th International Conference, Rovaniemi, 10–12 April*, pp. 51–66. Audio Engineering Society.

Berg, J. and Rumsey, F. (2000a). In search of the spatial dimensions of reproduced sound: verbal protocol analysis and cluster analysis of scaled verbal descriptors. Presented at *108th AES Convention, Paris, 19–22 February*. Preprint 5139. Audio Engineering Society.

Berg, J. and Rumsey, F. (2000b). Correlation between emotive, descriptive and naturalness attributes in subjective data relating to spatial sound reproduction. Presented at *109th AES Convention, Los Angeles, 22–25 September*. Preprint 5206.

Blauert, J. (1997). *Spatial Hearing. The Psychophysics of Human Sound Localisation*. MIT Press.

Bradley, J. and Souloudre, G. (1995). Objective measures of listener envelopment. *J. Acoust. Soc. Amer.*, **98**, pp. 2590–2597.

Bregman, A. (1990) *Auditory scene analysis: the perceptual organisation of sound*. MIT Press, Cambridge, Mass.

Gabrielsson, A. and Sjören, H. (1979). Perceived sound quality of sound reproducing systems. *J. Acoust. Soc. Amer.*, **65**, pp. 1019–1033.

Griesinger, D. (1997). Spatial impression and envelopment in small rooms. Presented at *103rd AES Convention, New York, 26–29 September*. Preprint 4638.

Griesinger, D. (1999). Objective measures of spaciousness and envelopment. In *Proceedings of the AES 16th International Conference, Rovaniemi, 10–12 April*, pp. 27–41. Audio Engineering Society.

Hafter, E. and Carrier, S. (1972). Binaural interaction in low frequency stimuli: the inability to trade time and intensity completely. *J. Acoust. Soc. Amer.*, **51**, pp. 1852–1862.

Harris, G. (1960). Binaural interactions of impulsive stimuli and pure tones. *J. Acoust. Soc. Amer.*, **32**, pp. 685–692.

Hebrank, J. and Wright, D. (1974). Spectral cues used in the localisation of sound sources on the median plane. *J. Acoust. Soc. Amer.*, **56**, pp. 1829–1834.

Howard, D. and Angus, J. (1996). *Acoustics and Psychoacoustics*. Focal Press, Oxford.

Huopaniemi, J. (1999). *Virtual acoustics and 3D sound in multimedia signal processing*. PhD thesis, Helsinki University of Technology.

Komiyama, S. (1989). Subjective evaluation of angular displacement between picture and sound directions for HDTV systems. *J. Audio Eng. Soc.*, **37**, 4, pp. 210–214.

Martin, G., Woszczyk, W., Corey, J. and Quesnel, R. (1999). Sound source localisation in a five channel surround sound reproduction system. Presented at *107th AES Convention, New York, 24–27 September*. Preprint 4994. Audio Engineering Society.

Mason, R. (1999). Personal communication.

Mason, R., Ford, N., Rumsey, F. and de Bruyn, B. (2000). Verbal and non-verbal elicitation techniques in the subjective assessment of spatial sound reproduction. Presented at *109th AES Convention, Los Angeles, 22–25 September*. Audio Engineering Society.

Mason, R. and Rumsey, F. (2000). An assessment of the spatial performance of virtual home theatre algorithms by subjective and objective methods. Presented at *108th AES Convention, Paris, 19–22 May*. Preprint 5137. Audio Engineering Society.

Moore, B. (1989). *An Introduction to the Psychology of Hearing*. Academic Press.

Nakayama, T. *et al.* (1971) Subjective assessment of multichannel reproduction. *J. Audio. Eng. Soc.*, **19**, pp. 744–751.

Nunally, J. and Bernstein, I. (1994). *Psychometric Theory*, 3rd ed. McGraw-Hill, New York and London.

Plenge, G. (1974). On the differences between localisation and lateralisation. *J. Acoust. Soc. Amer.*, **56**, pp. 944–951.

Ratliffe, P. (1974). Properties of hearing related to quadraphonic reproduction. *BBC Research Department Report, RD 1974/38*.

Simonsen, G. (1984). Master's thesis. Technical University of Lyngby, Denmark.

Tan, C-J. and Gan, W-S. (2000). Direct concha excitation for the introduction of individualised hearing cues. *J. Audio Eng. Soc.*, **48**, 7/8, pp.642–653.

Wallach, Newman and Rosenzweig (1949). The precedence effect in sound localisation. *Am. J. Psych.*, **62**, 3, p. 315.

Whitworth, R. and Jeffress, L. (1961). Time versus intensity in the localisation of tones. *J. Acoust. Soc. Amer.*, **33**, pp. 925–929.

Williams, M. (1987). Unified theory of microphone systems for stereophonic and sound recording. Presented at *82nd AES Convention*, London. Preprint 2466. Audio Engineering Society.

Yost, W., Wightman, F. and Green, D. (1971) Lateralisation of filtered clicks. *J. Acoust. Soc. Amer.*, **50**, pp. 1526–1531.

3 Two-channel stereo and binaural audio

This chapter describes the principles of conventional two-channel stereo as it relates to both loudspeaker and headphone reproduction. These principles are also useful when attempting to understand the capabilities and limitations of multichannel stereo as described in Chapter 4. The second part of this chapter deals with binaural audio – a means of representing three-dimensional sound scenes by encoding signals and reproducing them so as to feed the listener's ears with signals very similar to those that might be heard in natural listening.

3.1 Two-channel (2-0) stereo

Two-channel stereophonic reproduction (in international standard terms '2-0 stereo', meaning two front channels and no surround channels) is often called simply 'stereo' as it is the most common way that most people know of conveying some spatial content in sound recording and reproduction. In fact 'stereophony' refers to any sound system that conveys three-dimensional sound images, so it is used more generically in this book and includes surround sound. In international standards describing stereo loudspeaker configurations the nomenclature for the configuration is often in the form 'n-m stereo', where n is the number of front channels and m is the number of rear or side channels (the latter only being encountered in surround systems). This distinction can be helpful as it reinforces the slightly different role of the surround channels as explained in Chapter 4.

3.1.1 *Basic principles of loudspeaker stereo: 'Blumlein stereo'*

Based on a variety of formal research and practical experience, it has become almost universally accepted that the optimum configuration for two-loudspeaker stereo is an equilateral triangle with the listener located just to the rear of the point of the triangle (the loudspeaker forming the baseline). Beyond this, phantom images (the apparent locations of sound sources in-between the loudspeakers) become less stable, and the system is more susceptible to the effects of head rotation. This configuration gives rise to an angle subtended by the loudspeakers of ±30° at the listening position, as shown in Figure 3.1. In most cases stereo reproduction from two loudspeakers can only hope to achieve a modest illusion of three-dimensional spatiality, since reproduction is from the front quadrant only (although see Section 3.2.6 on 'transaural stereo'). That said, there has been a vast amount of research conducted on the basic 'stereophonic effect' and its optimisation, particularly for two-channel systems, as exemplified in the *AES Anthology on Stereophonic Techniques* (Eargle, 1986).

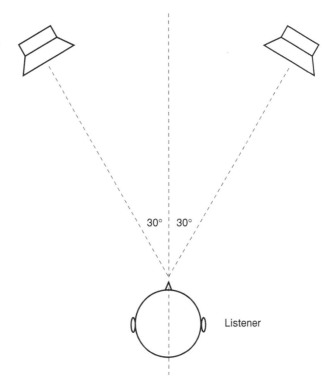

Figure 3.1 Optimum arrangement of two loudspeakers and listener for stereo listening.

30° 30°

Listener

Figure 3.2 An approximation to the situation that arises when listening to sound from two loudspeakers. Both ears hear sound from both loudspeakers, the signal from the right loudspeaker being delayed by δt at the left ear compared with the time it arrives at the right ear (and reversed for the other ear).

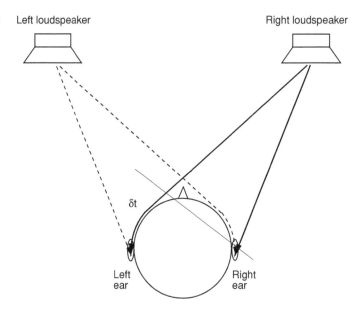

The so-called 'summing localisation' model of stereo reproduction suggests that the best illusion of phantom sources between the loudspeakers will be created when the sound signals present at the two ears are as similar as possible to those perceived in natural listening, or at least that a number of natural localisation cues that are non-contradictory are available. It is possible to create this illusion for sources in the angle between the loudspeakers using only amplitude differences between the loudspeakers, where the time difference between the signals is very small («1 ms). To reiterate an earlier point, in loudspeaker reproduction both ears receive the signals from both speakers, whereas in headphone listening each ear only receives one signal channel. The result of this is that the loudspeaker listener seated in a centre seat (see Figure 3.2) receives at his left ear the signal from the left speaker first followed by that from the right speaker, and at his right ear the signal from the right speaker first followed by that from the left speaker. The time ∂t is the time taken for the sound to travel the extra distance from the more distant speaker.

If the outputs of the two speakers differ only in amplitude and not in phase (time) then it can be shown (at least for low frequencies up to around 700 Hz) that the vector summation of the signals from the two speakers at each ear results in two signals which, for a given frequency, differ in phase angle proportional to the relative amplitudes of the two signals (the level difference between the ears being negligible at LF). For a given level differ-ence between the speakers, the phase angle changes approximately linearly with frequency, which is the case when listening to a real point source. At higher frequencies the phase difference cue becomes largely irrelevant but the shadowing effect of the head results in level differences between the ears. If the amplitudes of the two channels are correctly controlled it is possible to produce resultant phase and amplitude differences for continuous sounds that are very close to those experienced with natural sources, thus giving the impression of virtual or 'phantom' images anywhere between the left and right loudspeakers. This is the basis of Blumlein's 1931 stereophonic system 'invention' although the mathematics is quoted by Clark, Dutton and Vanderlyn (1958) and further analysed by others. The result of the mathematical phasor analysis is a simple formula that can be used to determine, for any angle subtended by the loudspeakers at the listener, what the apparent angle of the virtual image will be for a given difference between left and right levels. Firstly, referring to Figure 3.3, it can be shown that:

Figure 3.3 Real versus 'phantom' or virtual sound source location in stereo reproduction (see text).

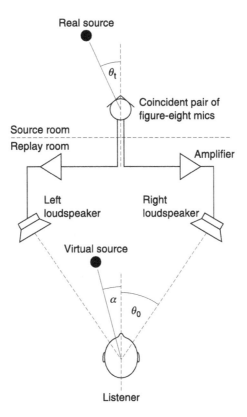

$$\sin \alpha = ((L - R)/(L + R)) \sin \vartheta_0$$

where α is the apparent angle of offset from the centre of the virtual image, and ϑ_0 is the angle subtended by the speaker at the listener. Secondly, it can be shown that:

$$(L - R)/(L + R) = \tan \vartheta_t$$

where ϑt is the true angle of offset of a real source from the centre-front of a coincident pair of figure-eight velocity microphones. $(L - R)$ and $(L + R)$ are the well-known difference (S) and sum (M) signals of a stereo pair, defined below.

This is a useful result since it shows that it is possible to use positioning techniques such as 'pan-potting' which rely on the splitting of a mono signal source into two components, with adjustment of the relative proportion fed to the left and right channels without affecting their relative timing. It also makes possible the combining of the two channels into mono without cancellations due to phase difference. Depending on which author one believes, an amplitude difference of between 15 and 18 dB between the channels is needed for a source to be panned either fully left or fully right. A useful summary of experimental data on this issue has been drawn by Hugonnet and Walder (1995), and is shown in Figure 3.4.

A coincident arrangement of velocity (figure-eight) microphones at 90° to one another produce outputs which differ in amplitude with varying angle over the frontal quadrant by an amount which theoretically gives a very close correlation between the true angle of offset of the original source from the centre line and the apparent angle on reproduction from loudspeakers which subtend an angle of 120° to the listening position (but this angle of loudspeakers is not found to be very satisfactory for practical purposes for reasons such as the tendency to give rise to a 'hole' in the middle of the image). At smaller loudspeaker angles the change in apparent angle is roughly proportionate as a fraction of total loudspeaker spacing, maintaining a correctly-

Figure 3.4 A summary of experimental data relating to amplitude differences (here labelled intensity) required between two loudspeaker signals for a particular phantom image location (data compiled by Hugonnet and Walder, 1995). Courtesy of Christian Hugonnet.

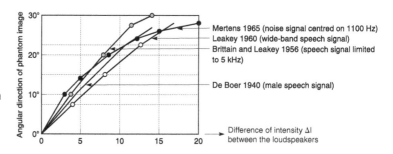

proportioned 'sound stage', so the sound stage with loudspeakers at the more typical 60° angle will tend to be narrower than the original sound stage but still in proportion. A number of people have pointed out that the difference in level between channels should be smaller at HF than at LF in order to preserve a constant relationship between actual and apparent angle, and this may be achieved by using a shelf equaliser in the difference channel (L − R) that attenuates the difference channel by a few dB for frequencies above 700 Hz (this being the subject of a British patent by Vanderlyn). Gerzon has suggested a figure between 4 and 8 dB (Gerzon, 1986) depending on programme material and spectral content, calling this 'spatial equalisation'. A similar concept has also been suggested by Griesinger (1986), except that in his approach he increased the LF width by increasing the gain in the difference channel below 300 Hz.

The ability of a system based only on level differences between channels to reproduce the correct timing of transient information at the ears has been questioned, not least in the discussion following the original paper presentation by Clark *et al.*, but these questions were tackled to some extent by Vanderlyn in a much later paper of 1979, in which he attempts to show how such a system can indeed result in timing differences between the neural discharges between the ears, taking into account the integrating effect of the hearing mechanism in the case of transient sounds. He quotes experimental evidence to support his hypothesis, which is convincing. If a system based only on level differences did not cope accurately with transients then one would expect transients to be poorly localised in subjective tests, and yet this is not the case, with transients being very clearly located in Blumlein-style recordings.

Summing localisation theories of two-channel stereo have been challenged by Theile in his paper 'On the naturalness of two-channel stereo sound' (Theile, 1991). Theile proposes an alternative model (the 'association model') that does not regard the reconstruction of ear signals as the most important factor, preferring to recommend the generation of 'head-referred' signals – that is signals similar to those used in binaural systems – since these contain the necessary information to reproduce a sound image in the 'simulation plane' between the two loudspeakers. He partly bases his model of perception on the observation that summing theories give rise to large spectral variations in the signals at the listener's ears as one moves or as a source is panned, but which are not directly perceived. He proposes that the brain is able to associate the spectral and timing differences between loudspeaker signals with the binaural encoding information that helps listeners to localise in natural hearing. Subjective tests have shown that binaural signals equalised for a flat frontal frequency response are capable of producing reasonably convincing stereo from loudspeakers, and there is still a degree of conflict between those adhering to this theory and those who adhere to the more conventional model of stereo reproduction. The issue of binaural audio on loudspeakers is discussed further later in this chapter, and microphone techniques adhering to this principle are described in Chapters 6 and 7.

3.1.2 Time-difference stereo

If a time difference also exists between the channels, then transient sounds will be 'pulled' towards the advanced speaker because of the precedence effect, the perceived position depending to some extent on the time delay. If the left speaker is advanced in time relative to the right speaker (or more correctly, the right speaker is delayed!) then the sound appears to come more from the left speaker, although this can be corrected by increasing the level to the right speaker. A delay somewhere between 0.5 and 1.5 ms is needed for a signal to appear fully left or fully right at ±30°, depending on the nature of the signal (see Figure 3.5, after Hugonnet and Walder). With time-difference stereo, continuous sounds may give rise to contradictory phantom image positions when compared with the position implied by transients, owing to the phase differences that are created between the channels. Cancellations may also arise at certain frequencies if the channels are summed to mono.

A trade-off is possible between interchannel time and level difference as described in principle in Chapter 2, although the exact relationship between time and level differences needed to place a source in a certain position is disputed by different authors and seems to depend to some extent on the source characteristics. (This inter-loudspeaker time–level trade-off must

Figure 3.5 A summary of experimental data relating to time differences required between two loudspeaker signals for a particular phantom image location (Hugonnet and Walder, 1995). Courtesy of Christian Hugonnet.

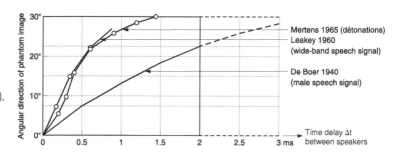

not be confused with binaural trading phenomena, as described in Chapter 2.) Stereo microphone techniques, as described in Chapter 6, operate using either interchannel level or time difference or a combination of the two.

3.1.3 Headphone reproduction compared to loudspeakers

Headphone reproduction is different to loudspeaker reproduction since, as already stated, each ear is fed only with one channel's signal. This is therefore an example of the binaural situation and allows for the ears to be fed with signals that differ in time by up to the binaural delay, and also differ in amplitude by amounts similar to those differences which result from the shadowing effects of the head. This suggests the need for a microphone technique which uses microphones spaced apart by the binaural distance, and baffled by an object similar to the human head, in order to produce signals with the correct differences. This is discussed in detail in the second part of this chapter.

Bauer (1961) pointed out that if stereo signals designed for reproduction on loudspeakers were fed to headphones there would be too great a level difference between the ears compared with the real life situation, and that the correct interaural delays would not exist. This results in an unnatural stereo image which does not have the expected sense of space and appears to be inside the head. He therefore proposed a network that introduced a measure of delayed crosstalk between the channels to simulate the correct interaural level differences at different

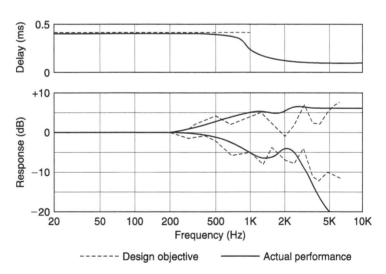

Figure 3.6 Bauer's filter for processing loudspeaker signals so that they could be reproduced on headphones. The upper graph shows the delay introduced into the crossfeed between channels. The lower graph shows the left and right channel gains needed to imitate the shadowing effect of the head.

frequencies, as well as simulating the interaural time delays which would result from loudspeaker signals incident at 45° to the listener. He based the characteristics on research done by Weiner which produced graphs for the effects of diffraction around the human head for different angles of incidence. The characteristics of Bauer's circuit are shown in Figure 3.6 (with Weiner's results shown dotted). It may be seen that Bauer chooses to reduce the delay at HF, partially because the circuit design would have been too complicated, and partially because localisation relies more on amplitude difference at HF anyway.

Interestingly Bauer's example of the stereophonic versus binaural problem chooses spaced pressure microphones as the means of pickup, showing that the output from the right microphone for signals at the left of the image will be near zero. This is only likely if the microphones are very close to the source (as in a multi microphone balance), whereas in many spaced omnidirectional arrays there will be a considerable output from the right microphone for sounds at the left of the sound stage, thus there will also be a time delay equivalent to the path length difference between the source and the two mikes which will add to that introduced by his network. In fact what Bauer is suggesting would probably work best with non-spaced microphones (i.e. a directional co-incident pair). It is a variation on Blumlein's 'shuffler' network. Blumlein's shuffler converted the phase differences between two binaurally-spaced microphones into amplitude variations to be reproduced correctly on loudspeakers, whereas Bauer is trying to insert a phase difference between two signals which differ only in level (as well as constructing a filter to simulate head diffraction effects).

Further work on a circuit for improving the stereo headphone sound image was done by Thomas (1977), quoting listening tests which showed that all of his listening panel preferred stereo signals on headphones which had been subjected to the 'cross-feed with delay' processing.

Bauer also suggests the reverse process (turning binaural signals into stereo signals for loudspeakers), pointing out that crosstalk must be removed between binaural channels for correct loudspeaker reproduction, since the crossfeed between the channels will otherwise occur twice (once between the pair of binaurally-spaced microphones, and again at the ears of the listener), resulting in poor separation and a narrow image. He suggests that this may be achieved using the subtraction of an anti-phase component of each channel from the other channel signal, although he does not discuss how the time difference between the binaural channels may be removed. Such processes are the basis of 'transaural stereo', introduced in Section 3.2.6.

3.1.4 Basic two-channel signal formats

The two channels of a 'stereo pair' represent the left (L) and the right (R) loudspeaker signals. It is conventional in broadcasting terminology to refer to the left channel of a stereo pair as the 'A' signal and the right channel as the 'B' signal, although this may cause confusion to some who use the term 'AB pair' to refer specifically to a spaced microphone pair. In the case of some stereo microphones or systems the left and right channels are called respectively the 'X' and the 'Y' signals, although some people reserve this convention specifically for coincident microphone pairs. Here we will stick to using L and R for simplicity. In colour coding terms (for meters, cables, etc.), particularly in broadcasting, the L signal is coloured red and the R signal is coloured green. This may be confusing when compared with some domestic hi-fi wiring conventions which use red for the right channel, but it is the same as the convention used for port and starboard on ships. Furthermore there is a German DIN convention which uses yellow for L and red for R.

It is sometimes convenient to work with stereo signals in the so-called 'sum and difference' format, since it allows for the control of image width and ambient signal balance. The sum or main signal is denoted 'M', and is based on the addition of L and R signals, whilst the difference or side signal is denoted 'S', and is based on the subtraction of R from L to obtain a signal which represents the difference between the two channels (see below). The M signal is that which would be heard by someone listening to a stereo programme in mono, and thus it is important in situations where the mono listener must be considered, such as in broadcasting. Colour-coding convention in broadcasting holds that M is coloured white, whilst S is coloured yellow, but it is sometimes difficult to distinguish between these two colours on certain meter types, leading to the increasing use of orange for S.

Two-channel stereo signals may be derived by many means. Most simply, they may be derived from a pair of coincident directional microphones orientated at a fixed angle to each other. Alternatively they may be derived from a pair of spaced microphones, either directional or non-directional, with an optional third microphone bridged between the left and right channels. Finally they may be derived by the splitting of one or more mono signals into two, by means of a 'pan-pot', which is really a dual ganged variable resistor which controls the relative proportion of the mono signal being fed to the two legs of the stereo pair, such that as the level to the left side is increased that to the right side is decreased. (The topic of recording techniques is covered in more detail in Chapter 6.)

MS or 'sum and difference' format signals may be derived by conversion from the LR format using a suitable matrix or by direct pickup in that format. For every stereo pair of signals it is possible to derive an MS equivalent, since M is the sum of L and R, whilst S is the difference between them. Likewise, signals may be converted from MS to LR formats using the reverse process. In order to convert an LR signal into MS format it is necessary to follow some simple rules. Firstly, the M signal is not usually a simple sum of L and R, as this will result in over-modulation of the M channel in the case where a maximum level signal exists on both L and R (representing a central image), therefore a correction factor is applied, ranging between –3 dB and –6 dB (equivalent to a division of the voltage by between √2 and 2 respectively), e.g.:

$$M = (L + R) - 3\,dB \text{ or } (L + R) - 6\,dB$$

The correction factor will depend on the nature of the two signals to be combined. If identical signals exist on the L and R channels (representing 'double mono' in effect), then the level of the uncorrected sum channel (M) will be two times (6 dB) higher than the levels of either of L or R, requiring a correction of –6 dB in the M channel in order for the maximum level of the M signal to be reduced to a comparable level. If the L and R signals are non-coherent (random phase relationship), then only a 3 dB rise in the level of M will result when L and R are summed, requiring the –3 dB correction factor to be applied. This is more likely with stereo music signals. As most stereo material has a degree of coherence between the channels, the actual rise in level of M compared with L and R is likely to be somewhere between the two limits for real programme material.

The S signal results from the subtraction of R from L, and is subject to the same correction factor, e.g.:

$$S = (L - R) - 3\,dB \text{ or } (L - R) - 6\,dB$$

S can be used to reconstruct L and R when matrixed in the correct way with the M signal (see below), since (M + S) = 2L and (M – S) = 2R. It may therefore be appreciated that it is possible at any time to convert a stereo signal from one format to the other and back again.

Figure 3.7 Two methods for MS matrixing. (a) Using transformers (passive method). (b) Using amplifiers (active method).

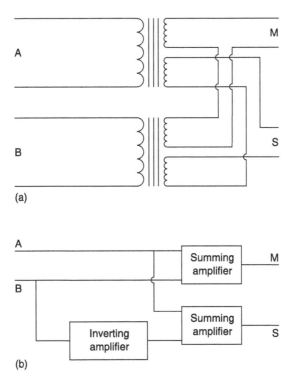

If either format is not provided at the output of microphones or mixing equipment it is a relatively simple matter to derive one from the other electrically. Figure 3.7 shows two possible methods. Figure 3.7(a) shows the use of transformers, where L and R signals are fed into the primaries, and M is derived by wiring the two secondaries in series and in phase, while S is derived by wiring two secondaries in series and out-of-phase. Figure 3.7(b) shows the use of summing amplifiers, whereby L and R are summed in-phase to derive M, and summed with an inversion in the R leg to derive S. Both matrixes have the advantage that they will also convert back to LR from MS, in that M and S may be connected to the inputs and they will be converted back to L and R.

3.1.5 Limitations of two-channel loudspeaker stereo

Two-channel stereo, without any form of special psychoacoustic processing, is limited in its ability to provide all-round sound images and reverberation. While it can sometimes give an illusion of this, and special panning circuits can create phantom images that extend a little beyond the loudspeakers, it is essentially limited to reproducing both sources and reverberation from an angle of about 60°. This is adequate for many purposes as the majority of listeners' attention is likely to be focused in front of them when listening to music or watching television. Phantom images are also subject to some tonal colouration as they are panned across the sound stage, owing to the way that the signals from two loudspeakers sum at the ears of the listener. A phantom central image will have a certain amount of mid-range colouration compared with that of an actual loudspeaker in that position. The lack of a centre loudspeaker also means that sound stages have a tendency to collapse into the nearest loudspeaker quite rapidly as one moves away from the ideal listening position or 'hot spot'. Various attempts have been made to compensate for this, for example by developing loudspeakers that have tailored directional characteristics that cover the far side of the listening area more strongly than the centre (so that the increased amplitude from the further speaker compensates for the precedence effect 'pull' of the nearer speaker).

3.2 Binaural sound and 3D audio systems

3.2.1 Introduction to binaural audio

As mentioned in Chapter 1, binaural recording has fascinated researchers for years but it has received very little commercial attention until recently. Part of the problem has been that it is actually very difficult to get it to work properly for a wide range of listeners over a wide range of different headphone types, and partly it is related to the limited compatibility between headphone and loudspeaker listening. Conventional loudspeaker stereo is acceptable on headphones to the majority of people, although it creates a strongly 'in-the-head' effect, but binaural recordings do not sound particularly good on loudspeakers without some signal processing and the stereo image is dubious. Record companies and broadcasters are unlikely to make two versions of a recording, one for headphones and one for loudspeakers, and equipment manufacturers have not seemed particularly interested in building conversion circuits into consumer equipment. Furthermore, dummy head recording, while interesting, has not been particularly good for creating the more 'commercial' sound that is desired by recording engineers, in which spot microphones and balance modifications are used.

Recent technical developments have made the signal processing needed to synthesise binaural signals and deal with the conversion between headphone and loudspeaker listening more widely available at reasonable cost. It is now possible to create 3D directional sound cues and to synthesise the acoustics of virtual environments quite accurately using digital signal processors (DSP), and it is this area of virtual environment simulation for computer applications that is receiving the most commercial attention for binaural technology today. Flight simulators, computer games, virtual reality applications and architectural auralisation are all areas that are benefiting from these developments. It is also possible to use such technology to synthesise 'loudspeakers' where none exist, using binaural cues, as employed in virtual home theatre systems (see below).

3.2.2 Basic binaural principles

Binaural approaches to spatial sound representation are based on the premise that the most accurate reproduction of natural spatial listening cues will be achieved if the ears of the listener can be provided with the same signals that they would have experienced in the source environment or during natural listening. Most of the approaches described so far in this chapter have related to loudspeaker reproduction of signals that contain *some* of the necessary information for the brain to localise phantom images and perceive a sense of spaciousness and depth. Much reproduced sound using loudspeakers relies on a combination of accurate spatial cues and believable illusion. In its purest form, binaural reproduction aims to reproduce *all* of the cues that are needed for accurate spatial perception, but in practice this is something of a tall order and various problems arise.

An obvious and somewhat crude approach to binaural audio is to place two microphones, one at the position of each ear in the source environment, and to reproduce these signals through headphones to the ears of a listener, as shown in Figure 3.8. Indeed this is the basic principle of binaural recording, and such a method can be used to striking effect, but there are numerous details that need to be considered before one can in fact recreate accurate spatial cues at the ears of the listener. Where exactly should the microphones be placed (e.g. in the pinna cavity, at the end of the ear canal)? What sort of ears and head should be used (everyone's is different)? What sort of headphones should be used and where should they be located (e.g. in the ear canal, on the pinna, open-backed, circumaural)? Can binaural signals be reproduced over loudspeakers? Some of the issues of headphone reproduction were covered earlier in Section 3.1.3, mainly concerning the differences between headphone and loudspeaker listening. Here we concentrate on the finer points

Figure 3.8 Basic binaural recording and reproduction.

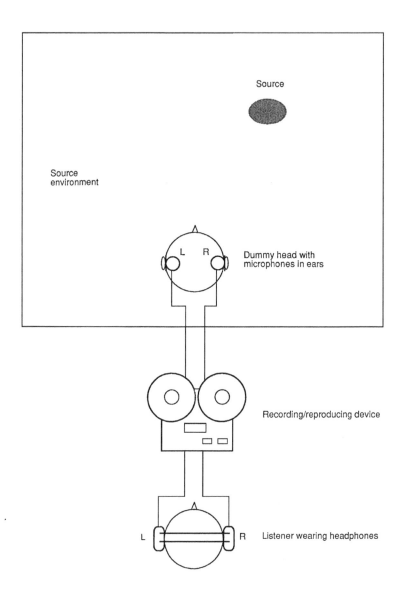

of how to ensure good correspondence between the recorded and reproduced spatial cues, and look at some of the technology that has made binaural systems more viable commercially.

For binaural reproduction to work well, the HRTFs of sound sources from the source (or synthesised) environment must be accurately recreated at the listener's ears upon reproduction (for a review of HRTF issues see Chapter 2). This means capturing the time and frequency spectrum differences between the two ears accurately. Since each source position results in a unique

HRTF, rather like a fingerprint, one might assume that all that is needed is to ensure the listener hears this correctly on reproduction.

3.2.3 Tackling the problems of binaural systems

The primary problems in achieving an accurate reconstruction of spatial cues can be summarised as follows:

- People's HRTFs are different (to varying degrees), although there are some common features, making it difficult to generalise about the HRTFs that should be used for commercial systems that have to serve lots of people.
- Head movements that help to resolve directional confusion in natural listening are difficult to incorporate in reproduction situations.
- Visual cues are often missing during binaural reproduction and these normally have a strong effect on perception.
- Headphones differ in their equalisation and method of mounting, leading to distortions in the perceived HRTFs on reproduction.
- Distortions such as phase and frequency response errors in the signal chain can affect the subtle cues required.

Begault reviewed a number of the challenges that were faced by those attempting to implement successful 3D sound systems based on binaural cues (Begault, 1991), and this is a useful paper for those wishing to study the matter further. He summarised the principal challenges for systems designers as (a) eliminating front-back reversals and intracranially heard sound (sounds inside the head); (b) reducing the amount of data needed to represent the most perceptually salient features of HRTF measurements; (c) resolving conflicts between desired frequency and phase response characteristics and measured HRTFs.

Considerable research has taken place in recent years to attempt to characterise the HRTFs of different subjects, and to create databases of features. To highlight the problem, Figure 3.9 shows 0° azimuth and elevation HRTFs for two subjects (from Begault's paper), showing just how much they can differ at high frequencies. Using methods such as principal components analysis and feature extraction it has been possible to identify the HRTF features that seem to occur in the majority of people and then to create generalised HRTFs that work reasonably well for a wide range of listeners. It has also been found that some people are better at localising sounds than others, and that the HRTFs of so-called 'good localisers' can be used in preference to those of

Figure 3.9 HRTFs of two subjects for a source at 0° azimuth and elevation. Note considerable HF differences. (Begault, 1991).

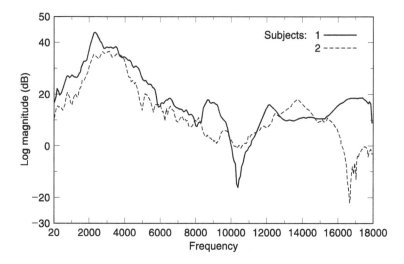

'poor localisers'. To summarise, it can be said that although a person's own HRTFs provide them with the most stable and reliable directional cues, generalised functions can be used at the expense of absolute accuracy of reproduction for everyone. There is evidence that subjects can gradually adapt to new HRTFs, such as those used in binaural sound systems, and that localisation errors become smaller with familiarity. Similarly, some adaption of the hearing process can take place when listening to reproduction with equalisation errors, many spatial effects being due more to *differences* between the signals at the two ears rather than the exact monaural HRTF characteristics.

Using tiny probe and miniature microphones, Møller *et al.* (1995a) measured the HRTFs of 40 subjects for seven directions of sound incidence. They claimed that measurements made at the entrance of the blocked ear canal characterised all the directional information and minimised individual differences between subjects. (A number of authors have claimed that the ear canal response has no directional encoding function, although there is not universal agreement on this.) Møller *et al.* observed that differences between subjects' HRTFs were relatively small up to about 8 kHz, and that above 8 kHz it was still possible to find a general structure for most directions. Averaging HRTFs across subjects was problematic as it tended to result in flattened curves that did not represent a typical subject, so more sophisticated techniques would be required to derive 'typical' or 'generic' functions.

The issue of head movements not affecting directional cues during binaural reproduction is dealt with in Section 3.2.5, and it can be helped by the use of head tracking. The issue of the lack of visual cues commonly encountered during reproduction can only be resolved in full 'virtual reality' systems that incorporate 3D visual information in addition to sound information. In the absence of visual cues, the listener must rely entirely on the sound cues to resolve things like front-back confusions and elevation/distance estimations.

The issue of headphone equalisation is a thorny one as it depends on the design goal for the headphones. Møller has suggested that for binaural reproduction the headphones should have or should be equalised to have a flat frequency response at the point in the ear where the binaural recording microphone was originally placed. In this way the spectral cues in the recording should be translated accurately to the reproduction environment. For emulating the timbre of sounds heard in loudspeaker reproduction, though, headphones have typically been equalised either to emulate the free-field response of a source at some angle in front of the listener (usually 0° but other angles have been proposed that correspond more nearly to a loudspeaker's position), or (as Theile and others suggest) to emulate the diffuse field response at the listening position (taking into account sound from all angles). (Real loudspeaker listening is usually some way in between the true free field and the completely diffuse field situations.)

When Møller *et al.* measured the responses at the entrance to the blocked ear canal of a variety of commercial headphones in 1995, they concluded that none of them seemed adequate for binaural reproduction without equalisation. Some of them came reasonably close to approximating the free or diffuse field design goal for reproduction of traditional loudspeaker stereo recordings. This rather suggests that it is close to impossible to devise a headphone that fulfils both the requirements for accurate binaural reproduction and accurate timbral matching to loudspeaker listening. Some form of switchable equalisation appears to be required, and preferably tailored to the individual. That said, various attempts have been made to equalise dummy heads and other binaural signals so that the differences between loudspeaker and headphone listening are not so great. Larcher *et al.* (1998) demonstrate that, for a variety of reasons, a diffuse field form of equalisation for headphones, dummy heads and synthesised environments is preferable to free-field equalisation. This is covered further in Section 3.2.4.

Distortions in the signal chain that can affect the timing and spectral information in binaural signals have been markedly reduced since the introduction of digital audio systems. In the days of analogue signal chains and media such as compact cassette and LP records, numerous opportunities existed for interchannel phase and frequency response errors to arise, making it difficult to transfer binaural signals with sufficient integrity for success.

3.2.4 Dummy heads, real heads and synthesised HRTFs

While it is possible to use a real human head for binaural recording (generally attached to a live person), it can be difficult to mount high quality microphones in the ears and the head movements and noises of the owner can be obtrusive. Sometimes heads are approximated by the use of a sphere or a disk separating a pair of microphones, and this simulates the shadowing effect of the head but it does not give rise to the other spectral filtering effects of the outer ear. Recordings made using such approaches have been found to have reasonable loudspeaker compatibility as they do not have the unusual equalisation that results from pinna filtering. (Unequalised true binaural recordings replayed on loudspeakers will typically suffer two stages of pinna filtering – once on recording and then again on reproduction – giving rise to distorted timbral characteristics.)

Dummy heads are models of human heads with pressure microphones in the ears that can be used for originating binaural signals suitable for measurement or reproduction. A number of commercial products exist, some of which also include either shoulders or a complete torso. A complete head-and-torso simulator is often referred to as a 'HATS', and an example is shown in Figure 3.10. The shoulders and torso are considered by some to be important owing to the reflections that result from them in natural listening, and which can contribute to the HRTF, but this has been found to be a factor that differs quite considerably between individuals and can therefore be a confusing cue if not well matched to the listener's own torso reflections.

Figure 3.10 Head and torso simulator (HATS) from B&K.

Some dummy heads or ear inserts are designed specifically for recording purposes whereas others are designed for measurement. As a rule, those designed for recording tend to have microphones at the entrances of the ear canals, whereas those designed for measurement have the mikes at the ends of the ear canals, where the eardrum should be. (Some measurement systems also include simulators for the transmission characteristics of the inner parts of the ear.) The latter types will therefore include the ear canal resonance in the HRTF, which would have to be equalised out for recording/reproduction purposes in which headphones were located outside the ear canal. The ears of dummy heads are often interchangeable in order to vary the type of ear to be simulated, and these ears are modelled on 'average' or 'typical' physical properties of human ears, giving rise to the same problems of HRTF standardisation as mentioned above.

The equalisation of dummy heads for recording has received much attention over the years, mainly to attempt better headphone/loudspeaker compatibility. Equalisation can be used to modify the absolute HRTFs of the dummy head in such a way that the overall spatial effect is not lost, partly because the differences between the ears are maintained. Just as Theile has suggested using diffuse field equalisation for headphones as a good means of standardising their response, he and others have also suggested diffuse field equalisation of dummy heads so that recordings made on such heads replay convincingly on such headphones and sound reasonably natural on loudspeakers. This essentially means equalising the dummy head microphone so that it has a near-flat response when measured in one-third octave bands in a diffuse sound field.

Gierlich and Genuit (1989), on the other hand, described an artificial head system (HEAD Acoustics) that was equalised for a flat 0° free-field response, claiming that this made the system most loudspeaker compatible. Griesinger (1989) experimented with various forms of equalisation for dummy heads and found that if (semi-)free or diffuse field measurements were averaged over one-third octave bands and the free field measurements were averaged over a 10° angle there was much less difference between the two for the angle of ±30° in front of the listener than had been suggested previously in the literature, which is encouraging. Such measurements may approximate more closely to the situation with music listening in real rooms. The Neumann KU100, pictured in Figure 3.11, is a dummy head that is designed to have good compatibility between loudspeaker and headphone reproduction, and uses equalisation that is close to Theile's proposed diffuse field response.

Figure 3.11 Neumann KU100 dummy head. Courtesy of Georg Neumann GmbH, Berlin.

Binaural cues do not have to be derived from dummy heads. Provided the HRTFs are known, or can be approximated for the required angle of sound incidence, signals can be synthesised with the appropriate time delays and spectral characteristics. Such techniques are increasingly used in digital signal processing applications that aim to simulate natural spatial cues, such as flight simulators and virtual reality. Accurate sets of HRTF data for all angles of incidence and elevation have been hard to come by until recently, and they are often quite closely guarded intellectual property as they can take a long time and a lot of trouble to measure. The question also arises as to how fine an angular resolution is required in the data set. For this reason a number of systems base their HRTF implementation on relatively coarse resolution data and interpolate the points in between.

3.2.5 Head tracking

Head tracking is a means by which the head movements of the listener can be monitored by the replay system. In some applications this information can be used to modify the binaural cues that are reproduced so that the head movements give rise to realistic changes in the signals sent to the ears. Generally this is only practical in real-time interactive applications where the HRTFs are updated continuously and individual virtual sources are being synthesised. The relationship between the direction in which the listener is facing and the intended virtual source locations can be calculated and the appropriate filter modifications effected. Experiments have indicated that the latency involved in the calculation of the filter resulting from a new head position can be reasonably long (<85 ms) without being perceived, as described in Section 3.2.8, although this may depend on the source characteristics and application. In this way, sources that are supposed to be at particular points in space relative to the listener can be made to stay in those places even if the listener moves. (In normal binaural reproduction the whole scene moves with the listener as the head is moved.)

Head tracking can help greatly in resolving front-back confusions, which are the bane of binaural reproduction. This was demonstrated by Begault (2000), where it was also shown that head tracking was not particularly important for improving the accuracy of source azimuth determination or externalisation of speech signals. It is often only by moving the head slightly that one can determine whether a source is in front or behind. Some head trackers also track movements of the head in the other two directions (that can be likened to tilt and yaw) in order that all three types of movement are taken into account.

Experiments conducted at the Institut für Rundfunkteknik, reported in Horbach *et al.* (1999), suggest that head tracking may be a crucial factor in improving the accuracy of binaural reproduction systems. In an experiment where a dummy head located in a separate room was motorised so that its movements could be made to track those of a listener's head they found that subjects' localisation of loudspeaker signals in the dummy head's room improved markedly with the head tracking turned on. Front-back reversals were virtually eliminated. Even more interestingly, they found that substituting the dummy head for a simple sphere microphone (no pinnae) produced very similar results, suggesting that the additional spectral cues provided by the pinnae were of relatively low importance compared with the effect of head rotation. It should be noted that the head and the headphones used in this experiment were both equalised for a flat diffuse field response, which makes the head similar to the sphere microphone in any case.

3.2.6 Replaying binaural signals on loudspeakers

When binaural signals are replayed on loudspeakers there is crosstalk between the signals at the two ears of the listener that does not occur with headphone reproduction (as shown earlier in Figure 3.2). The right ear gets the left channel signal a fraction of a second after it is received by the left ear, with an HRTF corresponding to the location of the left loudspeaker, and vice versa for the other ear. This prevents the correct binaural cues from being established at the listener's ears and eliminates the possibility for full 3D sound reproduction. Binaural stereo tends to sound excessively narrow at low frequencies when replayed on loudspeakers as there is very little difference between the channels that has any effect at a listener's ears. Furthermore, as mentioned above, the spectral characteristics of binaural recordings can create timbral inaccuracies when reproduced over loudspeakers unless some form of compromise equalisation is used.

The poor suitability of unprocessed binaural signals for loudspeaker reproduction has been challenged by Theile, as explained in Section 3.1.1, claiming that the brain is capable of associating 'head-related' differences between loudspeakers with appropriate spatial cues for stereo reproduction, provided the timbral quality of head-related signals is equalised for a natural-sounding spectrum (e.g. diffuse field equalisation, as described above). This theory has led to a variety of companies and recording engineers experimenting with the use of dummy heads such as the Neumann KU100 for generating loudspeaker signals, and spawned the idea for the Schoeps 'Sphere' microphone described in Chapter 6.

Griesinger (1989) proposed methods for the 'spatial equalisation' of binaural recordings to make them more suitable for loudspeaker reproduction. He suggested low frequency difference channel (L − R) boost of about 15 dB at 40 Hz (to increase

the LF width of the reproduction) coupled with overall equalisation for a flat frequency response in the total energy of the
recording to preserve timbral quality. This results in reasonably
successful stereo reproduction in front of the listener, but the
height and front–back cues are not preserved.

If the full 3D cues of the original binaural recording are to be
conveyed over loudspeakers, some additional processing is
required. If the left ear is to be presented only with the left
channel signal and the right ear with the right channel signal
then some means of removing the interaural crosstalk is
required. This is often referred to as crosstalk cancelling or
'transaural' processing. Put crudely, transaural crosstalk-
cancelling systems perform this task by feeding an anti-phase
version of the left channel's signal into the right channel and vice
versa, filtered and delayed according to the HRTF characteristic
representing the crosstalk path, as shown in Figure 3.12.

Figure 3.12 Basic principle
of a crosstalk cancelling
circuit.

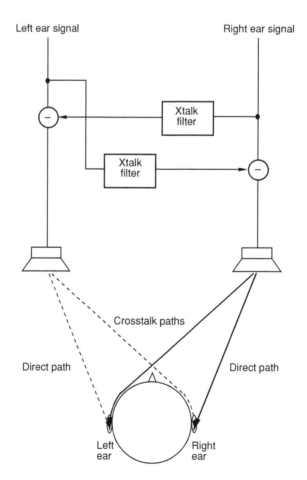

The effect of this technique can be quite striking, and in the best implementations enables fully three-dimensional virtual sources to be perceived, including behind the listener (from only two loudspeakers located at the front). The most important limitation is that the crosstalk-cancelling filters are only valid for a very narrow range of listening positions. Beyond a few tens of centimetres away from the 'hot spot' the effect often disappears almost completely. The effect is sometimes perceived as unnatural, and some listeners find it fatiguing to listen to for extended periods. As with most binaural systems, the engineering task facing most designers of recent years has been to find the optimum trade-off between localisation accuracy, compatibility with multiple listeners, timbral accuracy and robustness. Often one or other of these factors ends up suffering, so one finds systems that appear to be excellent for one or two people with their heads in a fixed position, but the image collapses completely as soon as they move and the result is not too good for other listeners, or one finds systems that work over a reasonably wide range of listening positions but are much more vague in their localisation accuracy and have timbral problems. Examples exist of quite good trade-offs of these factors, but the fact remains that this approach relies on listeners not being far off a known position, which is not satisfactory for many types of listening environment.

The situations in which the transaural approach has been most successful to date have been in 3D sound for desktop computers and virtual home theatre systems (see below). The reason for the considerable success in licensing the technology for computer sound cards is almost certainly that people operating desktop computers tend to sit in a highly predictable relationship with the display, to which can be attached the loudspeakers. This makes the filters easy to calculate and one doesn't need to allow for too much listener movement. Some systems are optimised for loudspeakers at ±5° positions for this very reason. Also, the sound quality and localisation accuracy required for computer games and other forms of multimedia entertainment are possibly not quite as exacting as for some other applications, and the loudspeakers are often of limited quality in any case, so one can opt for rather more crude and possibly exaggerated binaural cues in the system design.

3.2.7 Virtual surround or virtual home theatre (VHT) systems

VHT systems use binaural and transaural principles to 'virtualise' the surround loudspeakers of a 5.1-channel system, for environments in which it is not practical or desirable to have real loudspeakers. The centre channel is sometimes virtualised as well, although this can be dealt with as a simple phantom centre using conventional stereo techniques – the real challenge is getting sounds to appear to the sides and behind the listener. In such systems the LS and RS channels of the 5.1 surround mix are binaurally processed so as to create virtual sources with the HRTF corresponding to approximately 110° from the front on either side (the normal locations of the surround loudspeakers). The resulting signal is then fed through a transaural processor to cancel the interaural crosstalk as explained above, and the transaural signals are mixed with the front left and right channels of the 5.1 signal, as shown in Figure 3.13.

The subjective result can be reasonably convincing in the best examples, and quite unpleasant in the worst. Zacharov and Huopaniemi (1999) conducted a large-scale 'round-robin' test of a number of these systems, comparing the spatial and timbral quality with a hidden discrete 5.1-channel version of the same material. Not surprisingly, perhaps, the discrete version came out on top, with the VHT systems showing varying degrees of relative performance – some of them being consistently ranked very low in comparison. A primary problem noted was the severe timbral modification resulting from some processes. Such systems are typically encountered in some consumer televisions and surround replay equipment as they can be implemented in software and avoid the need to provide the extra physical outputs, speakers and amplifiers that would otherwise be required. They are often optimised for a reasonably wide listening area, so the effect deteriorates gradually as one moves away from the hot spot, and the resulting rear sound image is moderately diffuse (this is not normally a problem as the surround channels of most programme material are not intended to be accurately localised).

3.2.8 Headphone surround reproduction

There are situations in which one may wish to monitor loudspeaker surround programme material using headphones, and here binaural technology can come to the rescue again. Since headphones typically only have two transducers and we only have two ears, some means of mapping five or more loudspeaker signals into two ear signals has to be arranged.

Horbach *et al.* (1999) describe a headphone-based auralisation system for surround sound monitoring that virtualises the positions of the five loudspeakers and incorporates head-tracking

Figure 3.13 Virtualisation of surround and centre loudspeakers in virtual home theatre systems.

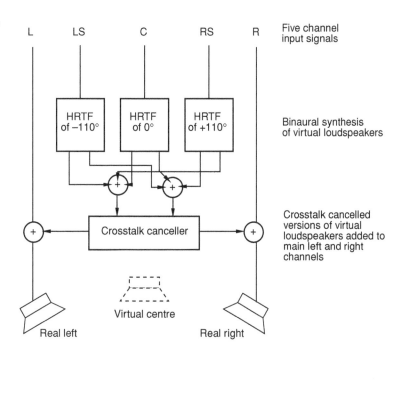

L   LS   C   RS   R   Five channel input signals

HRTF of −110°   HRTF of 0°   HRTF of +110°   Binaural synthesis of virtual loudspeakers

Crosstalk canceller   Crosstalk cancelled versions of virtual loudspeakers added to main left and right channels

Virtual centre

Real left   Real right

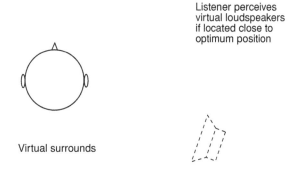

Listener perceives virtual loudspeakers if located close to optimum position

Virtual surrounds

so that the monitoring environment responds to head movements. In addition to the virtual loudspeaker signals, the system incorporates the real impulse responses of the acoustics of a sound control room, so that the loudspeakers sound as if they are playing in a natural environment. It is known that the addition of realistic reflections in the auralisation of sources contributes positively to a sense of externalisation of those sources.

The head tracking used in this system provides updates of head position every 8.3 ms, and the basic system latency is about

50 ms. Tests conducted by the authors to determine the effect of delays between head movements and the corresponding aural result of the filter update suggested that latency of less than 85 ms could not be detected by listeners.

3.2.9 Virtual acoustic environments

Binaural technology is highly suitable for the synthesis of virtual acoustic environments and is used increasingly in so-called 'auralisation' systems for acoustic design of real spaces, as well as for the creation of virtual spaces with their own acoustics.

Savioja *et al.* (1999) describe a number of techniques used for the modelling of acoustic spaces, and explains how they are implemented in the DIVA (digital virtual acoustics) system. The approach used separates the simulation of room acoustics into two parts as shown in Figure 3.14. Early reflections are simulated discretely using an image-source approach that is based on a real-time room acoustic model, updated according to the listener's position in the room and relative to the virtual source(s). Late reverberation that is naturally diffuse is not modelled in real time and can be pre-calculated from known room parameters. This makes the system computationally efficient. The basic structure of the system for filtering sources and early reflections according to their directional HRTFs, followed by the addition of reverberation and optional cross-talk cancelling for loudspeaker reproduction, is shown in Figure 3.15.

Figure 3.14 Two-part simulation of room acoustics used in virtual acoustic modelling (after Savioja *et al.*, 1999).

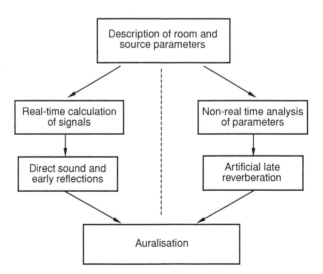

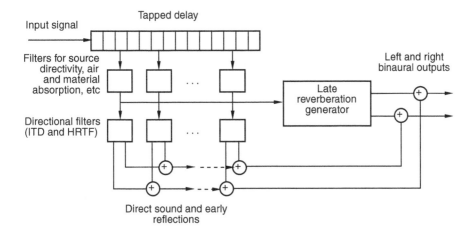

Input signal

Tapped delay

Filters for source
directivity, air
and material
absorption, etc

Directional filters
(ITD and HRTF)

Late
reverberation
generator

Left and right
binaural outputs

Direct sound and early
reflections

Figure 3.15 Outline of
signal processing elements
for virtual acoustic synthesis
(after Savioja *et al.*, 1999).

3.2.10 *Proprietary approaches to binaural reproduction*

A number of proprietary systems have been proposed over the years, relying in one way or another on binaural principles. Sometimes the nature of proprietary systems is cloaked in mystery, with claims that this or that black box will magically transform boring, flat sound images into beautiful three-dimensional sound spaces, but with little information about how this is achieved. Thankfully most such devices have been patented and patents are in the public domain if one has the tenacity to search for them.

Most notable was the Holophonic system, patented by Hugo Zucarelli in 1980. This aroused controversy at the time owing to the publicity surrounding the system and some unusual theories about spatial hearing propounded by Zucarelli. Investigation of the patent (European Patent 0 050 100) reveals that the system is little more than an accurately constructed dummy head, where the microphones are located at the eardrum ends of the ear canals and some equalisation is proposed to remove the effect of the ear canal resonance. A wig is proposed to make the head more realistic, and the oral cavity is simulated. Demonstrations at the time were regarded as very realistic, although they were often carefully stage-managed and used signals quite close to the head or moving (which typically give more exaggerated spatial cues). Holophonic effects were incorporated on a number of albums. An interesting article by Barry Fox in *Studio Sound* (Fox, 1983) looks into some of the history and debate surrounding Holophonics.

A system called *QSound* received a lot of attention in the late '80s and early '90s, being another system claiming to offer all-round sound source placement from two loudspeakers. It was licensed for use on a number of albums at the time. An investigation of the European Patent for QSound (No. 0 357 402) reveals that it is not in fact a system that relies on binaural synthesis techniques or crosstalk cancelling, but is based on empirically derived frequency-dependent phase and amplitude differences between the two channels of a stereo pair that give the impression of sources in different positions. While it is likely that these produce binaural signals at the listener's ears that correspond closely to natural source HRTFs for the virtual source locations concerned, no such analysis is provided.

A number of proprietary binaural/transaural systems have been developed more recently that rely strongly on digital signal processing for the synthesis of virtual sources using HRTF filtering. The intellectual property particular to each product generally seems to lie in the optimisation of HRTF details for different situations, computational efficiency and crosstalk cancelling algorithms. One example of such a system is *Sensaura*, originally developed by Central Research Labs in the UK. This is the result of a lot of research into the preservation of timbral accuracy in the signal chain from dummy head or synthesised sources, through crosstalk cancelling to the human listener. The system was designed to be particularly suitable for high quality music recording, although it has been licensed more recently for 3D audio features in consumer equipment and computers. The company developed a 3D audio workstation for mixing sound binaurally, providing signals that could be combined with the output of a dummy head.

References

Bauer, B. (1961). Phasor analysis of some stereophonic phenomena. *J. Acoust. Soc. Amer.* **33**, 11, pp. 1536–1539.

Begault, D. (1991). Challenges to the successful implementation of 3D sound. *J. Audio Eng. Soc.* **39**, 11, pp. 864–870.

Begault, D. (2000). Direct comparison of the impact of head tracking, reverberation, and individualized head-related transfer functions on the spatial perception of a virtual speech source. Presented at *AES 108th Convention, Paris, 19–22 February*. Preprint 5134. Audio Engineering Society.

Clark, H., Dutton, G. and Vanderlyn, P. (1958). The 'stereosonic' recording and reproducing system: a two-channel system for domestic tape records. *J. Audio Eng. Soc.*, **6**, 2, pp. 102–117.

Eargle, J. (ed.) (1986). *Stereophonic Technique: An Anthology of Reprinted Articles*. Audio Engineering Society.

Fox, B. (1983). Holophonics: an investigation. *Studio Sound*, July, pp. 90–96.

Gerzon, M. (1986). Stereo shuffling: new approach, old technique. *Studio Sound*, July.

Gierlich, H. and Genuit, K. (1989). Processing artificial head recordings. *J. Audio Eng. Soc.*, **37**, 1/2, pp. 34–39.

Griesinger, D. (1986). Spaciousness and localization in listening rooms and their effects on the recording technique. *J. Audio Eng. Soc.*, **34**, 4, pp. 255–268.

Griesinger, D. (1989). Equalization and Spatial Equalization of Dummy-Head Recordings for Loudspeaker Reproduction. *J. Audio Eng. Soc.*, **37**, 1/2, pp. 20–29.

Horbach, U. et al. (1999). Design and applications of a data-based auralization system for surround sound. Presented at *106th AES Convention, Munich, 8–11 May*. Preprint 4976.

Hugonnet, C. and Walder, P. (1995). *Stereophonic Sound Recording: Theory and Practice*. Wiley and Sons, Chichester.

Larcher, V., Jot, J-M. and Vandernoot, G. (1998). Equalisation methods in binaural technology. Presented at *AES 105th Convention, San Francisco, 26–29 September*, Preprint 4858. Audio Engineering Society.

Møller, H. et al. (1995a). Head-related transfer functions of human subjects. *J. Audio Eng. Soc.*, **43**, 5, pp. 300–321.

Møller, H. et al. (1995b). Transfer characteristics of headphones. *J. Audio Eng. Soc.*, **43**, 4, pp. 203–217.

Savioja, L., Huopaniemi, J., Lokki, T. and Väänänen, R. (1999). Creating interactive virtual acoustic environments. *J. Audio Eng. Soc.*, **47**, 9, pp. 675–705.

Theile, G. (1991). On the naturalness of two-channel stereo sound. *J. Audio Eng. Soc.*, **39**, 10, pp. 761–767.

Thomas, M. (1977). Improving the stereo headphone image. *J. Audio Eng. Soc.*, **25**, 7/8, pp. 474–478.

Zacharov, N. and Huopaniemi, J. (1999). Round-robin subjective evaluation of virtual home theatre sound systems at the AES 16th International Conference. In Proceedings of the *AES 16th International Conference, Rovaniemi, 10–12 April*, pp. 544–554. Audio Engineering Society.

4 Multichannel stereo and surround sound systems

This chapter is concerned with a description of the most commonly encountered multichannel (more than two channels) stereo reproduction configurations, most of which are often referred to as surround sound. There is an important distinction to be appreciated between standards or conventions that specify basic channel or speaker configurations and proprietary systems such as Dolby Digital and DTS whose primary function is the coding and delivery of multichannel audio signals. The latter are discussed in the second part of the chapter, in which is also contained an explanation of the Ambisonic system for stereo signal representation.

Surround sound standards often specify little more than the channel configuration and the way the loudspeakers should be arranged (e.g. 5.1-channel surround). This leaves the business of how to create or represent a spatial sound field entirely up to the user. To repeat a note from Chapter 3: in international standards describing stereo loudspeaker configurations the nomenclature for the configuration is often in the form '*n-m stereo*', where *n* is the number of front channels and *m* is the number of rear or side channels.

4.1 Three-channel (3-0) stereo

It is not proposed to say a great deal about the subject of three-channel stereo here, as it is rarely used in its own. Nonetheless

Figure 4.1 Three-channel stereo reproduction usually involves three equally spaced loudspeakers in front of the listener. The angle between the outer loudspeakers is 60° in the ITU standard configuration, for compatibility with two-channel reproduction, but the existence of a centre loudspeaker makes wider spacings feasible if compatibility is sacrificed.

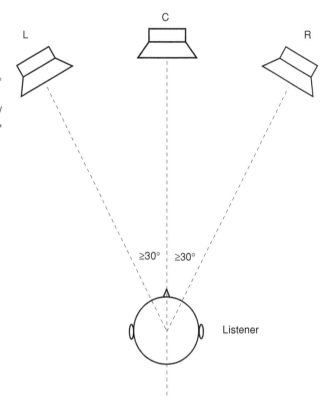

it does form the basis of a lot of surround sound systems. It requires the use of a left (L), centre (C) and right (R) channel, the loudspeakers arranged equidistantly across the front sound stage, as shown in Figure 4.1. It has some precedents in historical development, as mentioned in Chapter 1, in that the stereophonic system developed by Steinberg and Snow in the 1930s used three channels. Three front channels have also been commonplace in cinema stereo systems, mainly because of the need to cover a wide listening area and because wide screens tend to result in a large distance between left and right loudspeakers. Two channels only became the norm in consumer systems for reasons of economy and convenience, and particularly because it was much more straightforward to cut two channels onto an analogue disk than three.

Michael Gerzon started something of a campaign back in the early '90s to resurrect the idea of three-channel stereo (Gerzon, 1990), and published various circuits that could be used to derive a centre channel in a psychoacoustically optimum manner (Gerzon, 1992). Such circuits may be useful for the synthesis of a centre channel for 5.1 surround sound (see below) in a manner partially compatible with two-channel microphone techniques. He also devised panpot laws for three-channel stereo and showed how they compared with the original Bell Labs' laws (described in Chapter 6).

There are various advantages of three-channel stereo. Firstly it allows for a somewhat wider front sound stage than two-channel stereo, if desired, because the centre channel acts to 'anchor' the central image and the left and right loudspeakers can be placed further out to the sides (say ±45°). (Note, though, that in the current 5-channel surround sound standard the L and R loudspeakers are in fact placed at ±30°, for compatibility with two channel stereo.) Second, the centre loudspeaker enables a wider range of listening positions in many cases, as the image does not collapse quite as readily into the nearest loudspeaker. It also anchors dialogue more clearly in the middle of the screen in sound-for-picture applications. Thirdly, the centre image does not suffer the same timbral modification as the centre image in two-channel stereo, because it emanates from a real source.

A practical problem with three-channel stereo is that the centre loudspeaker position is often very inconvenient. Although in cinema reproduction it can be behind an acoustically transparent screen, in consumer environments, studios and television environments it is almost always just where one wants a television monitor or a window. Consequently the centre channel either has to be mounted above or below the object in question, and possibly made smaller than the other loudspeakers. This is covered further in Chapter 5.

4.2 Four-channel surround (3-1 stereo)

In this section the form of stereo called '3-1 stereo' in some international standards, or 'LCRS surround' in some other circles, is briefly described. Proprietary encoding and decoding technology from Dolby relating to this format is described later. 'Quadraphonic' reproduction using four loudspeakers in a square arrangement is not covered further here (it was mentioned in Chapter 1), as it has little relevance to current practice.

4.2.1 Purpose of 4-channel systems

The merits of three front channels have already been introduced in Section 4.1. In the 3-1 approach, an additional 'effects' channel or 'surround' channel is added to the three front channels, routed to a loudspeaker or loudspeakers located behind (and possibly to the sides) of listeners. It was developed first for cinema applications, enabling a greater degree of audience involvement in the viewing/listening experience by providing a channel for 'wrap-around' effects. Holman (1996) attributes this development to 20th Century Fox in the 1950s, along with widescreen Cinemascope viewing, being intended to offer effective competition to the new television entertainment.

There is no specific intention in 3-1 stereo to use the effects channel as a means of enabling 360° image localisation. In any case, this would be virtually impossible with most configurations as there is only a single audio channel feeding a larger number of surround loudspeakers, effectively in mono.

4.2.2 Loudspeaker configuration

Figure 4.2 shows the typical loudspeaker configuration for this format. In the cinema there are usually a large number of surround loudspeakers fed from the single S channel ('surround channel', not to be confused with the 'S' channel in sum-and-difference stereo), in order to cover a wide audience area. This has the tendency to create a relatively diffuse or distributed reproduction of the effects signal, and the speakers are sometimes electronically decorrelated to increase the degree of spaciousness or diffuseness in the surround channel, in order that the effects are not specifically localised to the nearest loudspeaker or perceived inside the head.

In consumer systems reproducing 3-1 stereo, the mono surround channel is normally fed to two surround loudspeakers located in similar positions to the 3-2 format described below. The gain of the channel is usually reduced by 3 dB so that the summation of signals from the two speakers does not lead to a level mismatch between front and rear.

4.2.3 Limitations of 4-channel reproduction

The mono surround channel is the main limitation in this format. Despite the use of multiple loudspeakers to reproduce the surround channel, it is still not possible to create a good sense of envelopment or spaciousness without using surround signals that are different on both sides of the listener. Most of the psychoacoustic research suggests that the ears need to be provided with decorrelated signals to create the best sense of envelopment and effects can be better spatialised using stereo surround channels. Proprietary systems, though, have used artificial decorrelation between surround loudspeakers driven

Figure 4.2 3-1 format reproduction uses a single surround channel usually routed (in cinema environments) to an array of loudspeakers to the sides and rear of the listening area. In consumer reproduction the mono surround channel may be reproduced through only two surround loudspeakers, possibly using artificial decorrelation and/or dipole loudspeakers to emulate the more diffused cinema experience.

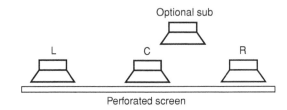

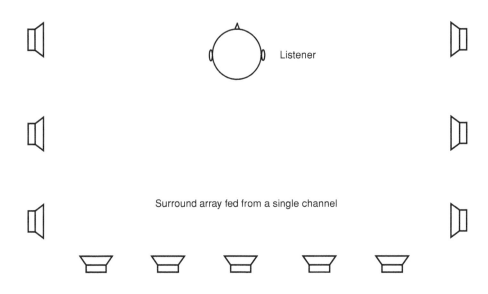

from a single channel to improve the spaciousness and diffusivity of the surround signal.

4.3 5.1-channel surround (3-2 stereo)

This section deals with the 3-2 configuration that has been standardised for numerous surround sound applications, including cinema, television and consumer applications. Because of its wide use in general parlance, the term '5.1 surround' will be used below. While without doubt a compromise, as mentioned in Chapter 1, it has become widely adopted in professional and consumer circles and is likely to form the basis for consumer surround sound for the foreseeable future.

Various international groups have worked on developing recommendations for common practice and standards in this area, and some of the information below is based on the effort of the AES Technical Committee on Multichannel and Binaural Audio Technology (led by the author) to bring together a number of proposals. European, Japanese and American contributions were incorporated.

4.3.1 Purpose of 5.1-channel systems

Four-channel systems have the disadvantage of a mono surround channel, and this limitation is removed in the 5.1-channel system, enabling the provision of stereo effects or room ambience to accompany a primarily front-orientated sound stage. This front-orientated paradigm is a most important one as it emphasises the intentions of those that finalised this configuration, and explains the insistence in some standards on the use of the term '3-2 stereo' rather than '5-channel surround'. Essentially the front three channels are intended to be used for a conventional three-channel stereo sound image, while the rear/side channels are only intended for generating supporting ambience, effects or 'room impression'. In this sense, the standard does not directly support the concept of 360° image localisation, although it may be possible to arrive at recording techniques or signal processing methods that achieve this to a degree (see Chapter 7).

The front–rear distinction is a conceptual point often not appreciated by those that use the format. While two-channel stereo can be relatively easily modelled and theoretically approached in terms of localisation vectors and such like, for sounds at any angle between the loudspeakers, it is more difficult to come up with such a model for the 5-channel layout described below, as it has unequal angles between the loudspeakers and a particularly large angle between the two rear loudspeakers. It is possible to arrive at gain and phase relationships between these five loudspeakers that are similar to those used in Ambisonics for representing different source angles, but the varied loudspeaker angles make the imaging stability less reliable in some sectors than others. For many using the format, who do not have access to the sophisticated panning laws or psychoacoustic matrices required to feed five channels accurately for all round localisation, it may be better to treat the format in 'cinema style', as one with a conventional (albeit three-channel) front image and two

Figure 4.3 3-2 format reproduction according to the ITU-R BS.775 standard uses two independent surround channels routed to one or more loudspeakers per channel.

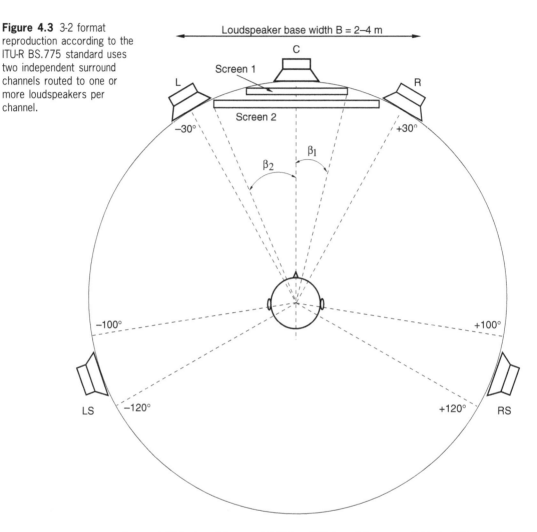

Screen 1: Listening distance = 3H (2 β_1 = 33°) (possibly more suitable for TV screen)
Screen 2: Listening distance = 2H (2 β_2 = 48°) (more suitable for projection screen)
H: Screen height

surround effect channels. With such an approach it is still possible to create very convincing spatial illusions, with good envelopment and localisation qualities.

One cannot introduce the 5.1 surround system without explaining the meaning of the '.1' component. This is a dedicated low frequency effects (LFE) channel or sub-bass channel, described in Section 4.3.4. It is called '.1' because of its limited bandwidth. Strictly, the international standard nomenclature for 5.1

surround should be '3-2-1', the last digit indicating the number of LFE channels.

4.3.2 International standards and configurations

The loudspeaker layout and channel configuration is specified in ITU-R BS.775 (ITU 1993). This is shown in Figure 4.3. A display screen is also shown in this diagram for sound with picture applications, and there are recommendations concerning the relative size of the screen and the loudspeaker base width shown in the accompanying table. The left and right loudspeakers are located at ±30° for compatibility with two-channel stereo reproduction. In many ways this need for compatibility with 2/0 is a pity, because the centre channel unavoidably narrows the front sound stage in many applications, and the front stage could otherwise take advantage of the wider dimension facilitated by three-channel reproduction. It was nonetheless considered crucial for the same loudspeaker configuration to be usable for all standard forms of stereo reproduction, for reasons most people will appreciate.

The surround loudspeaker locations, at approximately ±110°, are placed so as to provide a compromise between the need for effects panning behind the listener and the lateral energy important for good envelopment. In this respect they are more like 'side' loudspeakers than rear loudspeakers, and in many installations this is an inconvenient location causing people to mount them nearer the rear than the standard suggests. (Some have said that a 150° angle for the rear loudspeakers provides a more exciting surround effect.) In the 5.1 standard there are normally no loudspeakers directly behind the listener, which can make for creative difficulties. This has led to a Dolby proposal called EX (described below) that places an additional speaker at the centre-rear location. (This is not part of the current standard, though.) The ITU standard also allows for additional surround loudspeakers to cover the region around listeners, similar to the 3-1 arrangement described earlier. If these are used then they are expected to be distributed evenly in the angle between ±60° and ±150°.

Surround loudspeakers should be the same as front loudspeakers where possible, in order that uniform sound quality can be obtained all around. That said, there are arguments for the use of dipole loudspeakers in these positions. Dipoles radiate sound in more of a figure-eight pattern and one way of obtaining a diffuse surround impression is to orientate these with the nulls of the figure-eight towards the listening position. In this way the listener experiences more reflected than direct sound and this can give the impression of a more spacious ambient sound field that may better emulate the cinema listening experience in small rooms. Dipoles make it correspondingly more difficult to create defined sound images in rear and side positions, though. This issue is covered in greater detail in Chapter 5.

4.3.3 Track allocations and descriptions

Standards also recommend the track allocations to be used for 5.1 surround on eight-track recording formats, as shown in Table 4.1. Although other configurations are known to exist there is a strong move to standardise on this arrangement.

Table 4.1 Track allocations for 5.1 surround (based on recent international standards and proposals)

Track[1]	Signal		Comments	Colour[2]
1	L	Left		Yellow
2	R	Right		Red
3	C	Centre		Orange
4	LFE	Low frequency enhancement	Additional sub-bass and effects signal for subwoofer, optional[3]	Grey
5	LS	Left surround	−3 dB in the case of mono surround	Blue
6	RS	Right surround	−3 dB in the case of mono surround	Green
7	Free use in programme exchange[4]		Preferably left signal of a 2/0 stereo mix	Violet
8	Free use in programme exchange[4]		Preferably right signal of a 2/0 stereo mix	Brown

[1] The term 'track' is used to mean either tracks on magnetic tape, or virtual tracks on other storage media where no real tracks exist.

[2] This colour coding is only a proposal of the German Surround Sound Forum at present, and not internationally standardised.

[3] Preferably used in film sound, but is optional for home reproduction. If no LFE signal is being used, track 4 can be used freely, e.g. for commentary. In some regions a mono surround signal MS = LS + RS is applied, where the levels of LS and RS are decreased by 3 dB before summing.

[4] Tracks 7 and 8 can be used alternatively, for example, for commentary, for additional surround-signals, or for half-left/half-right front signal (e.g. for special film formats), or rather for the matrix format sum signal Lt/Rt.

4.3.4 The LFE channel and use of subwoofers

The low frequency effects channel is a separate sub-bass channel with an upper limit extending to a maximum of 120 Hz. It is intended to be used for conveying special low frequency content that requires greater sound pressure levels and headroom than can be handled by the main channels. It is not intended for conveying the low frequency component of the main channel signals, and its application is likely to be primarily in sound-for-picture applications where explosions and other high level rumbling noises are commonplace, although it may be used in other circumstances.

In consumer audio systems, reproduction of the LFE channel is considered optional. Because of this, recordings should normally be made so that they sound satisfactory even if the LFE channel is not reproduced. The EBU comments on the use of the LFE channel as follows:

> When an audio programme originally produced as a feature film for theatrical release is transferred to consumer media, the LFE channel is often derived from the dedicated theatrical subwoofer channel. In the cinema, the dedicated subwoofer channel is always reproduced, and thus film mixes may use the subwoofer channel to convey important low frequency programme content. When transferring programmes originally produced for the cinema over television media (e.g. DVD), it may be necessary to re-mix some of the content of the subwoofer channel into the main full bandwidth channels. It is important that any low frequency audio which is very significant to the integrity of the programme content is not placed into the LFE channel. The LFE channel should be reserved for extreme low frequency, and for very high level <120 Hz programme content which, if not reproduced, will not compromise the artistic integrity of the programme.

With cinema reproduction the in-band gain of this channel is usually 10 dB higher than that of the other individual channels. This is achieved by a level increase of the reproduction channel, not by an increased recording level. (This does not mean that the broadband or weighted SPL of the LFE loudspeaker should measure 10 dB higher than any of the other channels – in fact it will be considerably less than this as its bandwidth is narrower.)

It is a common misconception that any sub-bass or subwoofer loudspeaker(s) that may be used on reproduction must be fed directly from the LFE channel in all circumstances. While this may be the case in the cinema, bass management in the consumer reproducing system is not specified in the standard and is entirely system dependent. It is not mandatory to feed low frequency information to the LFE channel during the recording process, neither is it mandatory to use a subwoofer, indeed

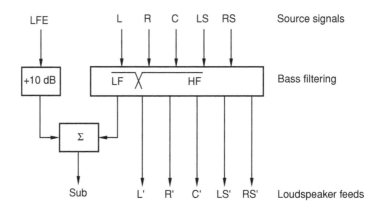

Figure 4.4 Low frequency management matrix for driving a sub-bass loudspeaker in a 5.1 reproduction system.

it has been suggested that restricting extreme low frequency information to a monophonic channel may limit the potential for low frequency spaciousness in balances. In music mixing it is likely to be common to send the majority of full range LF information to the main channels, in order to retain the stereo separation between them.

In practical systems it may be desirable to use one or more subwoofers to handle the low frequency content of a mix on reproduction. The benefit of this is that it enables the size of the main loudspeakers to be correspondingly reduced, which may be useful practically when it comes to finding places to put them in living rooms or sound control rooms. In such cases crossover systems split the signals between main loudspeakers and subwoofer(s) somewhere between 80 Hz and 160 Hz. In order to allow for reproduction of the LFE channel and/or the low frequency content from the main channels through subwoofer loudspeakers, a form of bass management akin to that shown in Figure 4.4 is typically employed. The issue of low frequency monitor setup is covered in more detail in Chapter 5.

4.3.5 Limitations of 5.1-channel reproduction

The main limitations of the 5.1 surround format are firstly that it was not intended for accurate 360° phantom imaging capability, as explained above. While it may be possible to achieve a degree of success in this respect, the loudspeaker layout is not ideally suited to it. Secondly, the front sound stage is narrower than it could be if compatibility with 2/0 reproduction was not a requirement. Thirdly, the centre channel can prove problematic for music balancing, as conventional panning laws and coincident microphone techniques are not currently optimised for three loudspeakers, having been designed for two-speaker stereo. Simple bridging of the centre loudspeaker between left and right signals has the effect of narrowing the front image compared with a two channel stereo reproduction of the same material. This may be resolved over time as techniques suited better to three-channel stereo are resurrected or developed, and some suggestions are made in Chapter 7. Fourthly, the LS and RS loudspeakers are located in a compromise position, leading to a large hole in the potential image behind the listener and making it difficult to find physical locations for the loudspeakers in practical rooms.

These various limitations of the format, particularly in some people's view for music purposes, have led to various non-standard uses of the five or six channels available on new consumer formats such as DVD-A and SACD. For example, some are using the sixth channel (that would otherwise be LFE) in its full bandwidth form on these media to create a height channel, whereas others are making a pair out of the 'LFE' channel and the centre channel so as to feed a pair of front-side loudspeakers, enabling the rear loudspeakers to be further back. These are non-standard uses and should be clearly indicated on any recordings.

4.3.6 *Signal levels in 5.1 surround*

Practice regarding alignment levels and maximum recording levels for 5.1 surround varies. In broadcasting and some studio recording operations, where programme interchange compatibility is of primary importance, it is normal to work to international standard guidelines that define an alignment level, L_{AS}, and a 'permitted maximum signal level', L_{PMS}. ITU and EBU recommendations, among others, specify a digital alignment signal level of -18 dBFS, whereas SMPTE recommendations specify -20 dBFS (1 kHz tone, RMS measurement). Both are likely to be encountered in operational practice, and it is therefore important to indicate clearly which alignment level is adopted, in order to avoid subsequent confusion.

The L_{PMS} is normally 9 dB below the digital clipping level, and is intended to be related to the measurement of programme signal on quasi-peak meters that have an integration time of 10 ms, thereby ensuring that short transients are not clipped. True peak reading meters will exceed this indication on some programme material, whereas VU meters will typically under-read this indication as they have a long integration time. In mastering and some film sound operations it is common to use the whole of the recording level range up to 0 dBFS, and in such circumstances it is important to use true peak-reading meters in order to avoid clipping on digital media.

In film sound environments it is the norm to increase the relative recording level of the surround channels by 3 dB compared with that of the front channels. This is in order to compensate for the –3 dB acoustic alignment of each surround channel's SPL with respect to the front that takes place in dubbing stages and movie theatres. It is important to be aware of this discrepancy between practices, as it is the norm in music mixing and broadcasting to align all channels for equal level both on recording media and for acoustical monitoring. Transfers from film masters to consumer or broadcast media may require 3 dB alteration in the gain of the surround channels.

4.4 Other multichannel configurations

Although the 5.1 surround standard is becoming widely adopted as the norm for the majority of installations, other proposals and systems exist, typically involving more channels to cover a large listening area more accurately. It is reasonable to assume that the more real loudspeakers exist in different locations around the listener the less one has to rely on the formation of phantom images to position sources accurately, and the more freedom one has in listener position. The added complication of mixing for such larger numbers of channels must be considered as a balancing factor.

The reader is also referred to the discussion of Ambisonics (Section 4.8), as this system can be used with a wide range of different loudspeaker configurations depending on the decoding arrangements used.

4.4.1 7.1-channel surround

Deriving from widescreen cinema formats, the 7.1-channel configuration normally adds two further loudspeakers to the 5.1-channel configuration, located at centre left (CL) and centre right (CR), as shown in Figure 4.5. This is not a format primarily intended for consumer applications, but for large cinema auditoria where the screen width is such that the additional channels are needed to cover the angles between the loudspeakers satisfactorily for all the seats in the auditorium. Sony's SDDS cinema system is a common proprietary implementation of this format, as is the original 70 mm Dolby Stereo format (see below), although the original 70 mm analogue format only used one surround channel.

Figure 4.5 Some cinema sound formats for large auditorium reproduction enhance the front imaging accuracy by the addition of two further loudspeakers, centre left and centre right.

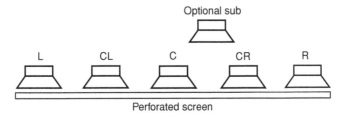

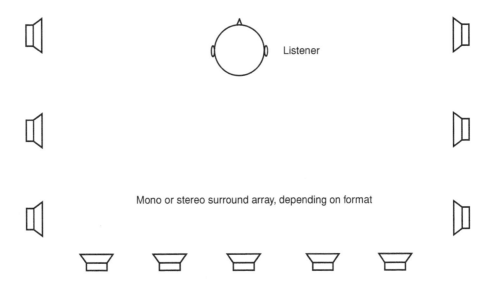

Lexicon has also implemented a 7-channel mode in its consumer surround decoder, but the recommended locations for the loudspeakers are not quite the same as in the cinema application, the additional channels being used to provide a wider side-front component and allow the rear speakers to be moved round more to the rear than in the 5.1 arrangement.

4.4.2 10.2-channel surround

Tomlinson Holman has spent considerable effort promoting a 10.2-channel surround sound system as 'the next step' in spatial reproduction, but this has not yet been adopted as a standard. To the basic 5-channel array he adds wider side-front loudspeakers and a centre-rear channel to 'fill in the holes' in the standard layout. He also adds two height channels and a second LFE channel. The second LFE channel is intended to provide lateral separation of decorrelated low bass content to either side of the listening area, as suggested by Griesinger to enhance low frequency spaciousness.

4.5 Surround sound systems

The second part of this chapter concerns what will be called surround sound 'systems', which includes proprietary formats for the coding and transfer of surround sound. These are distinguished from the generic principles of spatial reproduction and international standards discussed already. Most of the systems covered here are the subject of patents and intellectual property rights. In some proprietary systems the methods of signal coding or matrixing for storage and delivery are defined (e.g. Dolby Stereo), whilst others define a full source–receiver signal representation system (e.g. Ambisonics).

4.6 Matrixed surround sound systems

While ideally one would like to be able to transfer or store all the channels of a surround sound mix independently and discretely, it may be necessary to make use of existing two channel media for compatibility with other systems. The systems described in the following sections all attempt to deal with multichannel surround sound in a matrixed form. By matrixing the signals they can be represented using fewer channels than the source material contains. This gives rise to some side effects and the signals require careful dematrixing, but the approach has been used widely for many years, mainly because of the unavailability of multichannel delivery media in many environments.

4.6.1 Dolby Stereo, Surround and Prologic

Dolby Labs was closely involved with the development of cinema surround sound systems, and gradually moved into the area of surround sound for consumer applications.

The original Dolby Stereo system involved a number of different formats for film sound with three to six channels, particularly a 70 mm film format with six discrete tracks of magnetically recorded audio, and a 35 mm format with two optically recorded audio tracks onto which were matrixed four audio channels in the 3-1 configuration (described above). The 70 mm format involved L, LC, C, RC, R and S channels, whereas the 35 mm format involved only L, C, R and S. Both clearly only involved mono surround information. The four-channel system is the one most commonly known today as Dolby Stereo, having found widespread acceptance in the cinema world and used on numerous movies. Dolby Surround was introduced in 1982 as a means of emulating the effects of Dolby Stereo in a consumer environment. Essentially the same method of matrix decoding was used, so movies transferred to television formats could be decoded in the home in a similar way to the cinema. Dolby Stereo optical sound tracks for the cinema were Dolby A noise reduction encoded and decoded, in order to improve the signal to noise ratio, but this is not a feature of consumer Dolby Surround (more recent cinema formats have used Dolby SR-type noise reduction, alongside a digital soundtrack).

Figure 4.6 Basic components of the Dolby Stereo matrix encoding process.

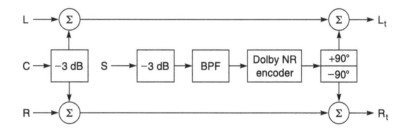

The Dolby Stereo matrix (see Figure 4.6) is a form of '4-2-4' matrix that encodes the mono surround channel so that it is added out of phase into the left and right channels (+90° in one channel and −90° in the other). The centre channel signal is added to left and right in phase. The resulting sum is called L_t/R_t (left total and right total). By doing this the surround signal can be separated from the front signals upon decoding by summing the L_t/R_t signals out of phase (extracting the stereo difference signal), and the centre channel can be extracted by summing L_t/R_t in phase. In consumer systems using passive decoding the centre channel is not always fed to a separate loudspeaker but can be heard as a phantom image between left and right. A decoder block diagram for the consumer version (Dolby Surround) is shown in Figure 4.7. Here it can be seen that in addition to the sum-and-difference-style decoding, the surround channel is subject to an additional delay, band-limiting between 100 Hz–7 kHz and a modified form of Dolby B noise reduction.

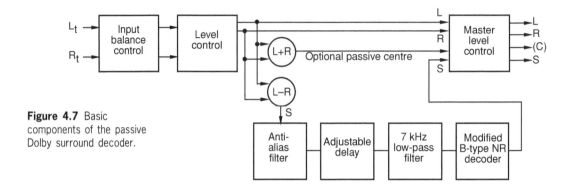

Figure 4.7 Basic components of the passive Dolby surround decoder.

The low pass filtering and the delay are both designed to reduce matrix side-effects that could otherwise result in front signals appearing to come from behind. Crosstalk between channels and effects of any misalignment in the system can cause front signals to 'bleed' into the surround channel, and this can be worse at high frequencies than low. The delay (of the order of 20–30 ms in consumer systems, depending on the distance of the rear speakers) relies on the precedence effect (see Chapter 2) to cause the listener to localise signals according to the first arriving wavefront which will now be from the front rather than the rear of the sound stage. The rear signal then becomes psychoacoustically better separated from the front and localisation of primary signals is biased more towards the front. The modified B-type NR reduces surround channel noise and also helps to reduce the effects of decoding errors and interchannel crosstalk, as some distortions introduced between encoding and decoding will be reduced by B-type decoding.

A problem with passive Dolby Surround decoding is that the separation between adjacent channels is relatively modest, although the separation of left/right and centre/surround remains high. When a signal is panned fully left it will tend to appear only 3 dB down in the centre, and also in the surround, for example. The effects of this can be ameliorated in passive consumer systems by the techniques described above (phantom centre and surround delay/filtering). Dolby's ProLogic system, based on principles employed in the professional decoder, attempts to resolve this problem by including sophisticated 'steering' mechanisms into the decoder circuit to improve the perceived separation between the channels. A basic block diagram is shown in Figure 4.8. This enables a real centre loudspeaker to be employed. Put crudely, ProLogic works by sensing the location of 'dominant' signal components and selectively attenuating channels away from the dominant component.

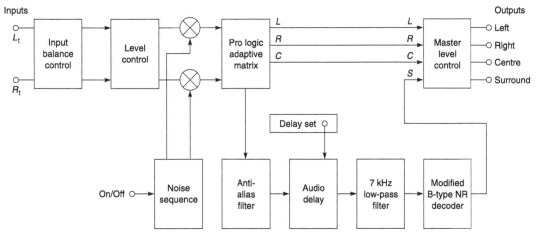

Figure 4.8 Basic components of the active Dolby Prologic decoder.

(A variety of other processes are involved as well as this.) So, for example, if a dialogue signal is predominantly located in the centre, the control circuit will reduce the output of the other channels (L, R, S) in order that the signal comes mainly from the centre loudspeaker (without this it would also have appeared at quite high level in left and right as well). A variety of algorithms are used to determine how quickly the system should react to changes in dominant signal position, and what to do when no signal appears dominant.

Mixes that are to be matrix encoded using the Dolby system should be monitored via the encode–decode chain in order that the side-effects of the process can be taken into account by the balance engineer. Dolby normally licenses the system for use on a project, and will assist in the configuration and alignment of their equipment during the project. The system works well for movie sound but is not particularly suited to music reproduction, because the stereo image tends to shift around as the programme content changes and the front image tends to be sucked towards the centre unless the original recording was mixed extremely wide. (Dolby is usually the first to acknowledge that the system was never designed for music reproduction purposes, although a number of people have experimented with mixing surround music recordings using Dolby Surround encoding with varying degrees of success.)

4.6.2 Circle Surround

Circle Surround was developed by the Rocktron Corporation (RSP Technologies) as a matrix surround system capable of encoding stereo surround channels in addition to the conventional front channels. They proposed the system as more appropriate than Dolby Surround for music applications, and claimed that it should be suitable for use on material that had not been encoded as well as that which had.

The Circle Surround encoder is essentially a sum and difference Lt/Rt process (similar to Dolby but without the band limiting and NR encoding of the surround channel), one incarnation of which involves 5-2 encoding, intended for decoding back to five channels (the original white paper on the system described a 4-2 encoder). Among other methods, the Circle decoder steers the rear channels separately according to a split-band technique that steers low and high frequency components independently from each other. In this way they claim to avoid the broad-band 'pumping' effects associated with some other systems. They also decode the rear channels slightly differently, using L–R for the left rear channel and R–L for the right rear channel, which it is claimed allows side images to be created on either side. They avoid the use of a delay in the rear channels for the 'Music' mode of the system and do not band-limit the rear channels as Dolby Surround does.

In the Music mode, a signal panned hard left will normally appear in the left rear channel as well, with a 3 dB gain increase. This is said to be to allow for producers to use conventional stereo panning to locate sounds around the listener. In the author's experience it also has the effect of putting an uncomfortably high level of signal in the surround channels when decoding conventional two-channel material that was not mixed specifically for Circle Surround, causing the original front image to be distributed in an unusual fashion around the listener. This reinforces the importance of monitoring surround mixes intended for matrix encoding through the complete encode-decode chain, and to check for conventional two-channel compatibility.

There is a 'Video' mode as well, which is claimed to improve upon Dolby decoding by retaining a wider front sound stage (by attenuating the centre channel dynamically when there is not an obvious centre component) and allowing the rear channels to be steered in a manner similar to the Music mode, or alternatively emulating the perceived characteristics of the Dolby decoder.

4.6.3 Lexicon Logic 7

Logic 7 is another surround matrix decoding process that can be used as an alternative for Dolby Surround decoding. Variants on

Figure 4.9 Approximate loudspeaker layout suitable for Lexicon's Logic 7 reproduction. Notice the additional side loudspeakers that enable a more enveloping image and may enable rear loudspeakers to be placed further to the rear.

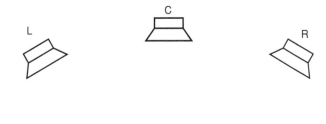

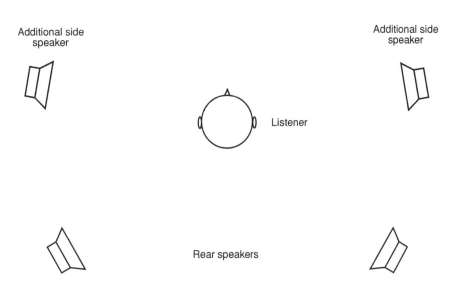

Additional side speaker

Additional side speaker

Listener

Rear speakers

this algorithm (such as the so-called Music Logic and Music Surround modes) can also be used for generating a good surround effect from ordinary two-channel material. Lexicon developed the algorithm for its high-end consumer equipment, and it is one of a family of steered decoding processes that distributes sound energy appropriately between a number of loudspeakers depending on the gain and phase relationships in the source material. In this case seven loudspeaker feeds are provided rather than five, adding two 'side' loudspeakers to the array, as shown in Figure 4.9. The rear speakers can then be further to the rear than would otherwise be desirable. The side loudspeakers can be used for creating an enhanced envelopment effect in music modes and more accurate side panning of effects in movie sound decoding.

In Logic 7 decoding of Dolby matrix material the front channel decoding is almost identical to Dolby ProLogic, with the addition of a variable centre channel delay to compensate for non-ideal locations of the centre speaker. The rear channels operate differently depending on whether the front channel content is primarily steered dialogue or effects (in which case these signals are cancelled from the rear channels and panned effects behave as they would with ProLogic, with surround effects panned 'full rear' appearing in mono on both rear channels), or music/ambience (in which case the rear channels work in stereo, but reproducing the front left and right channels with special equalisation and delay to create an enveloping spatial effect). The side channels carry steered information that attempts to ensure that effects which pan from left to rear pass through the left-side on the way, and similarly for the right side with right-to-rear pans.

It is claimed that by using these techniques the effect of decoding a 3-1 matrix surround version of a 3-2 format movie can be brought close to that of the original 3-2 version. Matrix encoding of five channels to Lt/Rt is also possible with a separate algorithm, suitable for decoding to five or more loudspeakers using Logic 7.

4.6.4 Dolby EX

In 1998 Dolby and Lucasfilm THX joined forces to promote an enhanced surround system that added a centre rear channel to the standard 5.1-channel setup. They introduced it, apparently, because of frustrations felt by sound designers for movies in not being able to pan sounds properly to the rear of the listener – the surround effect typically being rather diffuse. This system was christened 'Dolby Digital – Surround EX', and apparently uses matrix-style centre channel encoding and decoding between the left and right surround channels of a 5.1-channel mix. The loudspeakers at the rear of the auditorium are then driven separately from those on the left and right sides, using the feed from this new 'rear centre' channel, as shown in Figure 4.10.

4.7 Digital surround sound formats

Matrix surround processes are gradually giving way to digital formats that enable multiple channels to be delivered discretely, bypassing the two-channel restriction of most previous delivery formats. While it is desirable to be able to store or transfer multichannel surround sound signals in a discrete, full-resolution digital PCM format, this can occupy considerable amounts of storage space or transmission bandwidth (somewhere between about 0.75 and 2 Mbit/s per channel, depending on the resolution). This is too high for practical purposes in broadcasting, film sound and consumer systems, using current technology.

Figure 4.10 Dolby EX adds a centre-rear channel fed from a matrix-decoded signal that was originally encoded between left and right surround channels in a manner similar to the conventional Dolby Stereo matrix process.

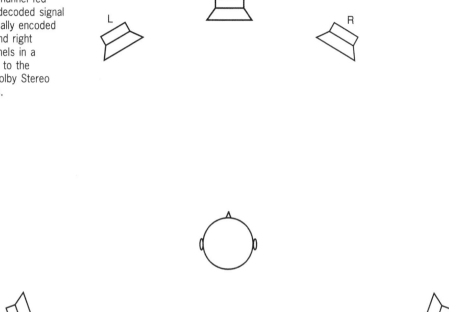

Matrix derived
rear centre speaker

Consequently a number of approaches have been developed whereby the information can be digitally encoded at a much lower bit rate than the source material, with minimal loss of sound quality. The sections below briefly describe some of these systems, used in cinema sound, digital consumer formats and broadcasting systems.

4.7.1 Dolby Digital

Dolby Digital or AC-3 encoding was developed as a means of delivering 5.1-channel surround to cinemas or the home without

103

Figure 4.11 Dolby Digital data is stored optically in an area between the sprocket holes of 35 mm film. (Courtesy of Dolby Laboratories.)

the need for analogue matrix encoding. It is likely to replace Dolby Stereo/Surround gradually as digital systems replace analogue ones. It relies on a digital low-bit-rate encoding and decoding process that enables the multiple channels of the surround mix to be conveyed without the separation and steering problems inherent in matrixed surround. Dolby Digital can code signals based on the ITU-standard 3-2-1 surround format of loudspeakers and it should be distinguished from such international standards since it is primarily a signal coding and representation method. In fact, the AC-3 coding algorithm can be used for a wide range of different audio signal configurations and bit rates from 32 kbit/s for a single mono channel up to 640 kbit/s for surround signals. It is used widely for the distribution of digital sound tracks on 35 mm movie films, the data being stored optically in the space between the sprocket holes on the film, as shown in Figure 4.11. In this way, the analogue optical sound tracks can be retained in their normal place alongside the picture for compatibility purposes. In this format it is combined with a Dolby-SR encoded analogue Dolby Stereo mix, and the combined format is called Dolby SR-D. Dolby Digital is also used for surround sound on DVD video releases, and for certain digital broadcasting applications.

The principles of Dolby Digital encoding and decoding are not really relevant to the purposes of this book, being more suited to a book on data rate reduction. The interested reader is referred to a paper by Craig Todd and colleagues for further information (Todd *et al.*, 1994). It is sufficient to say here that the process involves a number of techniques by which the data representing audio from the six source channels is transformed into the frequency domain and requantised to a lower resolution, relying on the masking characteristics of the human hearing process to hide the increased quantising noise that results from this process. A common bit pool is used so that channels requiring higher data rates than others can trade their bit rate requirements provided that the overall total bit rate does not exceed the

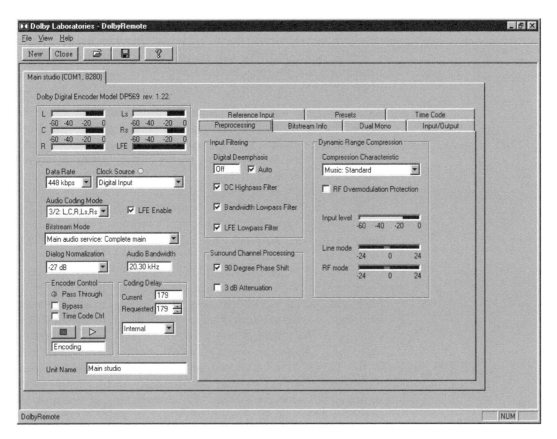

Figure 4.12 Screen display of Dolby Digital encoding software options.

constant rate specified. The result is a single bitstream that contains all the information for the six channels at a typical data rate of around 320–384 kbit/s (a reduction of over ten times in the raw PCM data rate), although a very wide range of bit rates and channel configurations is possible in theory. Some small differences in sound quality can sometimes be noticed between the original uncompressed PCM data and the version that has been encoded and decoded, but these are designed to be minimal and are the result of a compromise between sound quality and data rate. If a higher data rate is used the sound quality can be made higher.

The Dolby Digital encoding process can be controlled by a software application that enables various parameters of the encoding process to be varied, as shown in Figure 4.12. Dolby Digital can operate at sampling rates of 32, 44.1 or 48 kHz, and the LFE channel is sampled at 240 Hz (because of its limited bandwidth). A 90° phase shift is normally introduced into each of the surround channels during encoding, which apparently improves the smoothness of front–back panning and reduces crosstalk between centre and surround channels when decoded to Dolby Surround. For this reason it is important to monitor recordings via the encode–decode process to ensure that this phase shift does not affect the spatial intention of the producer.

Aside from the representation of surround sound in a compact digital form, Dolby Digital includes a variety of operational features that enhance system flexibility and help adapt replay to a variety of consumer situations. These include dialogue normalisation ('dialnorm') and the option to include dynamic range control information alongside the audio data for use in environments where background noise prevents the full dynamic range of the source material to be heard. Downmix control information can also be carried alongside the audio data in order that a two-channel version of the surround sound material can be reconstructed in the decoder. Downmixing is covered further in Chapter 7, but here it is simply mentioned that the Dolby Digital bitstream contains information that allows the mastering engineer or original sound balancer to generate the downmix coefficients that control the process, so that the downmix is reasonably satisfactory from an artistic point of view. As a rule, Dolby Digital data is stored or transmitted with the highest number of channels needed for the end product to be represented, and any compatible downmixes are created in the decoder. This differs from some other systems where a two-channel downmix is carried alongside the surround information.

Dialnorm indication can be used on broadcast and other material to ensure that the dialogue level remains roughly constant from programme to programme (it is assumed that this is the main factor governing the listening level used in people's homes, and that they do not like to have to keep changing this as different programmes come on the air (e.g. from advertising to news programmes). The dialnorm level is the average dialogue level over the duration of the programme compared to the maximum level that would be possible, measured using an A-weighted L_{EQ} reading (this averages the level linearly over time). So, for example, if the dialogue level averaged at 70 dBA over the programme, and the SPL corresponding to peak recording level was 100 dBA, the dialnorm setting would be –30 dB.

4.7.2 DTS

The DTS (Digital Theater Systems) 'Coherent Acoustics' system is another digital signal coding format that can be used to deliver surround sound in consumer or professional applications, using low bit rate coding techniques to reduce the data rate of the audio information. The DTS system can accommodate a wide range of bit rates from 32 kbit/s up to 4.096 Mbit/s (somewhat higher than Dolby Digital), with up to eight source channels and with sampling rates up to 192 kHz. Variable bit rate and lossless coding are also optional. Downmixing and dynamic range control options are provided in the system. The principles are outlined in a paper by Smyth *et al.* (1996).

DTS data is found on some film releases and occupies a different area of the film to Dolby Digital and SDDS data (see below). In fact it is possible to have film release prints in a multi-format version with all three digital sound formats plus the analogue Dolby Stereo tracks on one piece of film, making it almost universally compatible. DTS is also used on a number of surround CD releases and is optional on DVD, requiring a special decoder to replay the data signal from the disk. Because the maximum data rate is typically somewhat higher than that of Dolby Digital or MPEG, a greater margin can be engineered between the signal and any artefacts of low bit rate coding, leading to potentially higher sound quality. Such judgements, though, are obviously up to the individual and it is impossible to make blanket statements about comparative sound quality between systems.

4.7.3 SDDS

SDDS stands for Sony Dynamic Digital Sound, and is the third of the main competing formats for digital film sound. Using Sony's ATRAC data reduction system, it too encodes audio data with a substantial saving in bit rate compared with the original PCM (about 5:1 compression). The SDDS system employs 7.1 channels rather than 5.1, as described earlier in this chapter, providing detailed positional coverage of the front sound stage. It is not common to find SDDS data on anything but film release prints, although it could be included on DVD as a proprietary format if required. Consumer decoders are not currently available, to the author's knowledge.

4.7.4 MPEG

The MPEG (Moving Pictures Expert Group) standards are widely used for low bit rate representation of audio and video signals in multimedia and other applications. While MPEG-1 described a two-channel format, MPEG-2 extended this to multi-channel information. There are two versions of MPEG-2, one of which was developed to be backwards compatible with MPEG-1 decoders and the other of which is known as MPEG-2 AAC (for advanced audio coding) and is not backwards compatible. The MPEG-4 standards also include scalable options for multichannel coding. The standards are described in detail in ISO 11172-3, 13818-3, 13818-7 and 14496 for those who want to understand how they work in detail.

The MPEG-2 BC (backwards compatible) version worked by encoding a matrixed downmix of the surround channels and the centre channel into the left and right channels of an MPEG-1 compatible frame structure. This could be decoded by conventional MPEG-1 decoders. A multichannel extension part was then added to the end of the frame, containing only the C, LS and RS signal channels, as shown in Figure 4.13. Upon decoding in an MPEG-2 surround decoder, the three additional surround components could be subtracted again from the L_0/R_0 signals to leave the original five channels. The main problems with MPEG-2 BC are that (a) the downmix is performed in the encoder so it cannot be changed at the decoder end, and (b) the data rate required to transfer the signal is considerably higher than it would be if backward compatibility were not an issue. Consequently the bit rate required for MPEG-2 BC to transfer 5.1-channel surround with reasonable quality is in the region of 600–900 kbit/s.

MPEG-2 AAC, on the other hand, is a more sophisticated algorithm that codes multichannel audio to create a single bit stream that represents all the channels, in a form that cannot be decoded by an MPEG-1 decoder. Having dropped the requirement for backward compatibility, the bit rate can now be optimised by coding the channels as a group and taking advantage of interchannel redundancy if required. The situation is now more akin to that with Dolby Digital, and the bit rates required for acceptable sound quality are also similar. The MPEG-2 AAC system contained contributions from a wide range of different manufacturers, and the concept is described well by Bosi *et al.* (1997).

The MPEG surround algorithms have not been widely implemented to date in broadcasting, film and consumer applications. Although MPEG-2 BC was originally intended for use with DVD releases in Region 2 countries (primarily Europe), this requirement appears to have been dropped in favour of Dolby Digital. MPEG two-channel standards, such as MPEG-1, Layer 3 (the well known .MP3 format), have been widely adopted for consumer purposes on the other hand.

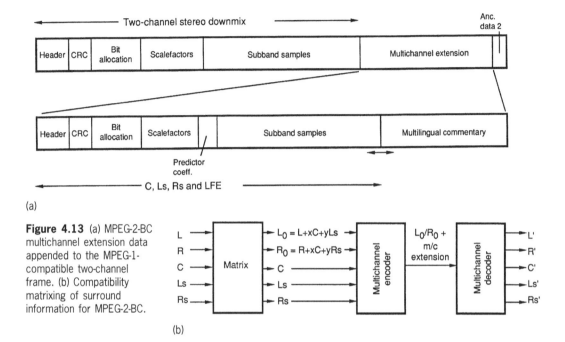

Two-channel stereo downmix

Anc. data 2

| Header | CRC | Bit allocation | Scalefactors | Subband samples | Multichannel extension | |

| Header | CRC | Bit allocation | Scalefactors | | Subband samples | Multilingual commentary |

Predictor coeff.

C, Ls, Rs and LFE

(a)

Figure 4.13 (a) MPEG-2-BC multichannel extension data appended to the MPEG-1-compatible two-channel frame. (b) Compatibility matrixing of surround information for MPEG-2-BC.

$L_0 = L + xC + yLs$
$R_0 = R + xC + yRs$

$L_0/R_0 + $ m/c extension

(b)

4.7.5 *MLP*

Meridian Lossless Packing (MLP) is a lossless data reduction technique for multichannel audio, licensed by Meridian Audio through Dolby Labs. It has been specified for the DVD-Audio format as a way of reducing the data rate required for high quality recordings without any effect on sound quality (in other words, you get back exactly the same bits you put in, which is not the case with lossy processes like Dolby Digital and MPEG). Using this technique, a sufficient playing time can be obtained from the disk whilst still enabling high audio resolution (sample rate up to 192 kHz and resolution between 16 and 24 bits) and up to six-channel surround sound.

MLP enables the mastering engineer to create a sophisticated downmix (for two-channel replay) of the multichannel material that occupies very little extra space on the disk, owing to the exploitation of similarities between this material and the multi-channel version during lossless encoding. This downmix can have characteristics that vary during the programme and is entirely under the artistic control of the engineer.

There are also modes of MLP that have not really seen the light of day yet. For example, the system is extensible to considerable numbers of channels, and has an option to incorporate hierarchical encoding processes such as Ambisonics where sound field components rather than loudspeaker feeds are represented. This could be useful in future as a means of overcoming the limitations of a loudspeaker-feed-based format for delivering surround sound to consumers.

4.7.6 THX

THX is not a digital surround sound delivery system, but is described at the end of this section on proprietary systems as it is designed to enhance a number of aspects of surround sound reproduction.

The THX system was developed by Tomlinson Holman at Lucasfilm (THX is derived from 'Tomlinson Holman Experiment'). The primary aim of the system was to improve the sound quality in movie theatres and make it closer to the sound experienced by sound mixers during post-production. It was designed to complement the Dolby Stereo system, and does not itself deal with the encoding or representation of surround sound. In fact THX is more concerned with the acoustics of cinemas and the design of loudspeaker systems, optimising the acoustic characteristics and noise levels of the theatre, as well as licensing a particular form of loudspeaker system and crossover network. THX licenses the system to theatres and requires that the installation is periodically tested to ensure that it continues to meet the specification.

Home THX was developed, rather like Dolby Surround, in an attempt to convey the cinema experience to the home. Through the use of a specific controller, amplifiers and speakers, the THX system enhances the decoding of Dolby Surround and can also be used with digital surround sound signals. The mono surround signal of Dolby Surround is subject to decorrelation of the signals sent to the two surround loudspeakers in order that the surround signal is made more diffuse and less 'mono'. It is claimed that this has the effect of preventing surround signals from collapsing into the nearest loudspeaker. Signals are re-equalised to compensate for the excessive high frequency content that can arise when cinema balances are replayed in small rooms, and the channels are 'timbre matched' to compensate for the spectral changes that arise when sounds are panned to different positions around the head (see Chapter 2).

In terms of hardware requirements, the Home THX system also specifies certain aspects of amplifier performance, as well as controlling the vertical and horizontal directivity of the front loudspeakers (vertical directivity is tightly controlled to increase the direct sound component arriving at listeners, while horizontal directivity is designed to cover a reasonably wide listening area). Front speakers should have a frequency response from 80 Hz to 20 kHz and all speakers must be capable of radiating an SPL of 105 dB without deterioration in their response or physical characteristics. The surround speakers are unusual in having a bipolar radiation pattern, arranged so that the listener hears reflected sound rather than direct sound from these units. These have a more relaxed frequency response requirement of 125 Hz to 8 kHz. A subwoofer feed is usually also provided.

4.8 Ambisonics

4.8.1 Principles

The Ambisonic system of directional sound pickup and reproduction is discussed here because of its relative thoroughness as a unified system, being based on some key principles of psychoacoustics. It has its theoretical basis in work by Gerzon, Barton and Fellgett, good summaries of which may be found in Gerzon (1973, 1974, 1977). It also has its origin in work undertaken earlier by Cooper and Shiga (1972).

Ambisonics aims to offer a complete hierarchical approach to directional sound pickup, storage or transmission and reproduction, which is equally applicable to mono, stereo, horizontal surround-sound, or full 'periphonic' reproduction including height information. Depending on the number of channels employed it is possible to represent a lesser or greater number of dimensions in the reproduced sound. A number of formats exist for signals in the ambisonic system, and these are as follows: the A-format for microphone pickup, the B-format for studio equipment and processing, the C-format for transmission, and the D-format for decoding and reproduction. A format known as UHJ ('Universal HJ', 'HJ' simply being the letters denoting two earlier surround sound systems), described originally by Gaskell of the BBC Research Department (Gaskell, 1979), is also used for encoding multichannel ambisonic information into two or three channels whilst retaining good mono and stereo compatibility for 'non-surround' listeners.

Ambisonic sound should be distinguished from quadraphonic sound, since quadraphonics explicitly requires the use of four loudspeaker channels, and cannot be adapted to the wide variety of pickup and listening situations which may be encountered.

111

Quadraphonics generally works by creating conventional stereo phantom images between each pair of speakers and, as Gerzon states, conventional stereo does not perform well when the listener is off-centre or when the loudspeakers subtend an angle larger than 60°. Since in quadraphonic reproduction the loudspeakers are angled at roughly 90° there is a tendency towards a hole-in-the-middle, as well as there being the problem that conventional stereo theories do not apply correctly for speaker pairs to the side of the listener. Ambisonics however, encodes sounds from all directions in terms of pressure and velocity components, and decodes these signals to a number of loudspeakers, with psychoacoustically optimised shelf filtering above 700 Hz to correct for the shadowing effects of the head and an amplitude matrix which determines the correct levels for each speaker for the layout chosen. Ambisonics might thus be considered as the theoretical successor to coincident stereo on two loudspeakers, since it is the logical extension of Blumlein's principles to surround sound.

The source of an ambisonic signal may be an ambisonic microphone such as the Calrec Soundfield, described in Chapter 7, or it may be an artificially panned mono signal, split into the correct B-format components (see below) and placed in a position around the listener by adjusting the ratios between the signals. Good introductions to the subject of mixing may be found in Daubney (1982) and Elen (1983).

It is often the case that theoretical 'correctness' in a system does not automatically lead to widespread commercial adoption, and despite considerable coverage of ambisonic techniques the system is still only used rarely in commercial recording and broadcasting. It is true that the Soundfield microphone is used quite widely, but this is principally because of its unusual capacity for being steered electrically so as to allow the microphone to be 'pointed' in virtually any direction without physically moving it, and set to any polar pattern between omni and figure-eight, simply by turning knobs on a control box. It is used in this respect as an extremely versatile stereo microphone for two-channel recording and reproduction.

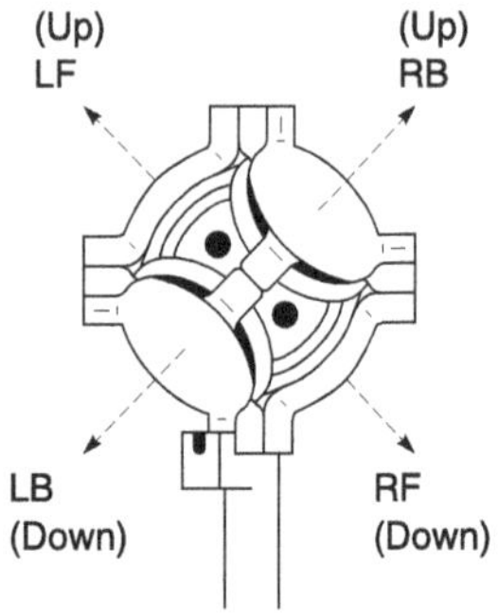

4.8.2 Signal formats

As indicated above there are four basic signal formats for ambisonic sound: A, B, C and D. The A-format consists of the four signals from a microphone with four sub-cardioid capsules orientated as shown in Figure 4.14 (or the panpot equivalent of such signals). These are capsules mounted on the four faces of a

Figure 4.14 A-format capsule directions in an Ambisonic microphone.

Figure 4.15 The B-format components W, X, Y and Z in Ambisonics represent an omnidirectional pressure component and three orthogonal velocity (figure-eight) components of the sound field respectively.

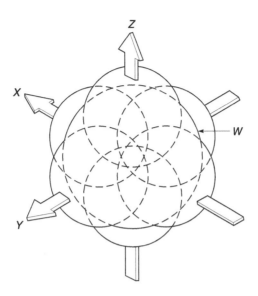

tetrahedron, and correspond to left-front (LF), right-front (RF), left-back (LB) and right-back (RB), although two of the capsules point upwards and two point downwards. Such signals should be equalised so as to represent the soundfield at the centre of the tetrahedron, since the capsules will not be perfectly coincident. The A-format is covered further in the discussion of the Soundfield microphone in Chapter 7.

The B-format consists of four signals that between them represent the pressure and velocity components of the sound field in any direction, as shown in Figure 4.15. It can be seen that there is a similarity with the sum and difference format of two channel stereo, described earlier, since the B-format is made up of three orthogonal figure-eight components (X, Y and Z), and an omni component (W). All directions in the horizontal plane may be represented by scalar and vector combinations of W, X and Y, whilst Z is required for height information. X is equivalent to a forward-facing figure-eight (equivalent to M in MS stereo), Y is equivalent to a sideways-facing figure-eight (equivalent to S in MS stereo). The X, Y and Z components have a frontal, sideways or upwards gain of +3 dB or √2 with relation to the W signal (0 dB) in order to achieve roughly similar energy responses for sources in different positions.

A B-format signal may be derived from an A-format microphone. In order to derive B-format signals from these capsule-pair signals it is a simple matter of using sum and difference technique. Thus:

$$X = 0.5((LF - LB) + (RF - RB))$$

$$Y = 0.5((LF - RB) - (RF - LB))$$

$$Z = 0.5((LF - LB) + (RB - RF))$$

W, being an omni pressure component, is simply derived by adding the outputs of the four capsules in phase, thus:

$$W = 0.5(LF + LB + RF + RB)$$

In a microphone W, X, Y and Z are corrected electrically for the differences in level between them, so as to compensate for the differences between pressure and velocity components. For example, W is boosted at very low frequencies since it is derived from velocity capsules which do not have the traditionally extended bass response of omnis.

B-format signals may also be created directly by arranging capsules or individual microphones in the B-format mode (two or three figure-eights at 90° plus an omni). The Z component is not necessary for horizontal information. If B-format signals are recorded instead of speaker feeds (D-format), subsequent manipulation of the soundfield is possible, and the signal will be somewhat more robust to interchannel errors.

Taking ϑ as the angle of incidence in the horizontal plane (the azimuth), and η as the angle of elevation above the horizontal, then in the B-format the polar patterns of the different signals can be represented as follows:

$$W = 1$$

$$X = \sqrt{2} \cos\vartheta \, \cos\eta$$

$$Y = \sqrt{2} \sin\vartheta \, \cos\eta$$

$$Z = \sqrt{2} \sin\eta$$

The C-format consist of four signals L, R, T and Q, which conform to the UHJ hierarchy, and are the signals used for mono or stereo-compatible transmission or recording. The C-format is, in effect, a useful consumer matrix format. L is a two-channel-compatible left channel, R is the corresponding right channel, T is a third channel which allows more accurate horizontal decoding, and Q is a fourth channel containing height information. The proportions of B-format signals which are combined to make up a C-format signal have been carefully optimised for the best compatibility with conventional stereo and mono reproduction. If L + R is defined as Σ (similar to M in MS stereo) and L − R is defined as Δ (similar to S in MS stereo), then:

$$\Sigma = 0.9397W + 0.1856X$$

$$\Delta = j(-0.3420W + 0.5099X) + 0.6555Y$$

$$T = j(-0.1432W + 0.6512X) - 0.7071Y$$

$$Q = 0.9772Z$$

where j (or $\sqrt{-1}$) represents a phase advance of 90°.

Two, three, or four channels of the C-format signal may be used depending on the degree of directional resolution required, with a two-and-a-half channel option available where the third channel (T) is of limited bandwidth. For stereo compatibility only L and R are used (L and R being respectively $0.5(\Sigma + \Delta)$ and $0.5(\Sigma - \Delta)$). The UHJ or C-format hierarchy is depicted graphically in Figure 4.16.

D-format signals are those distributed to loudspeakers for reproduction, and are adjusted depending on the selected loudspeaker layout. They may be derived from either B- or C-format signals using an appropriate decoder, and the number of speakers is not limited in theory, nor is the layout constrained to a square. Four speakers give adequate surround sound, whilst six provide better immunity against the drawing of transient and sibilant signals towards a particular speaker, whilst eight may be used for full periphony with height. The decoding of B- and C-format components into loudspeaker signals is too complicated and lengthy a matter to go into here, and is the subject of

Figure 4.16 The C-format or UHJ hierarchy enables a variety of matrix encoding forms for stereo signals, depending on the amount of spatial information to be conveyed and the number of channels available.

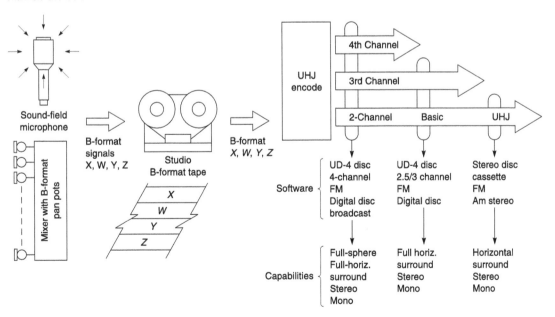

Figure 4.17 C-format signals are decoded to provide D-format signals for loudspeaker reproduction.

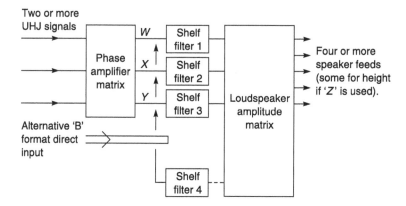

several patents that were granted to the NRDC (the UK National Research and Development Council, as was), but it is sufficient to say that the principle of decoding involves the passing of two or more UHJ signals via a phase-amplitude matrix, resulting in B-format signals that are subjected to shelf filters (in order to correct the levels for head-related transfer functions such as shadowing and diffraction). These are passed through an amplitude matrix which feeds the loudspeakers (see Figure 4.17). A layout control is used to vary the level sent to each speaker depending on the physical arrangement of speakers. Formulae relating to a number of decoding options may be found in Gerzon (1983).

Ambisonic encoding and decoding is also reviewed in an excellent paper by Jot, Larcher and Pernaux (1999). In this paper they note David Malham's observation that Ambisonic decoding can give rise to antiphase components from diametrically opposed speaker pairs in an array. While this may be desirable to optimise the localisation vectors at an ideal listening position it is found to be far from ideal for large auditorium listening or for situations where many listeners are away from the ideal listening position, or where loudspeaker locations are not ideal and possibly subject to errors. They propose various alternative decoding strategies that deal with different requirements such as the need to minimise antiphase components.

4.8.3 Higher order Ambisonics

The incorporation of additional directional components into the Ambisonic signal structure can give rise to improved directional encoding that covers a larger listening area than first order Ambisonics. These second order and higher components are part of a family of so-called 'spherical harmonics'. Horizontal Ambisonics can be enhanced by the addition of two further components, U and V, which have polar patterns described by:

$$U = \sqrt{2}\cos(2\vartheta)$$

$$V = \sqrt{2}\sin(2\vartheta)$$

provided that an appropriate decoder is implemented that can deal with the second order components. Even higher order components can be generated with the general form:

$$c_n \text{ (forwards)} = \sqrt{2}\cos(n\vartheta)$$

$$c_n \text{ (sideways)} = \sqrt{2}\sin(n\vartheta)$$

The problem with higher order Ambisonics is that it is much more difficult to design microphones that produce the required polar patterns, although the signals can be synthesised artificially for sound modelling and rendering applications.

4.8.4 B-format-to-5.1 decoding

Although the original ambisonic specifications assumed symmetrical rectangular or square loudspeaker layouts, Gerzon showed in 1992 how ambisonic signals could be decoded with reasonable success to layouts such as the 5-channel configuration described above (Gerzon and Barton, 1992). These are often referred to as 'Vienna decoders' after the location of the AES Convention at which these were first described. The sound image is in this case 'front biased', with better localisation characteristics in the frontal region than the rear, owing to the loudspeaker layout. This is an unavoidable feature of such a configuration in any case.

References

Bosi, M. *et al.* (1997). ISO/IEC MPEG-2 advanced audio coding. *J. Audio. Eng. Soc.*, **45**, pp. 789–812.

Cooper, D. and Shiga, T. (1972). Discrete matrix multichannel stereo. *J. Audio. Eng. Soc.*, **20**, pp. 346–360.

Daubney, C. (1982). Ambisonics – an operational insight. *Studio Sound*, August, pp. 52–58.

Elen, R. (1983). Ambisonic mixing – an introduction. *Studio Sound*, September, pp. 40–46.

Gaskell, P. (1979). System UHJ: a hierarchy of surround sound transmission systems. *The Radio and Electronics Engineer*, **49**, pp. 449–459.

Gerzon, M. (1973). Periphony: with-height sound reproduction. *J. Audio. Eng. Soc.*, **21**, pp. 2–10.

Gerzon, M. (1974). Surround sound psychoacoustics. *Wireless World*, **80**, pp. 483–486.

Gerzon, M. (1977). Criteria for evaluating surround sound systems. *J. Audio. Eng. Soc.*, **25**, pp. 400–408.

Gerzon, M. (1983) Ambisonics in multichannel broadcasting and video. Presented at *74th AES Convention, New York, October*. Preprint 2034. Audio Engineering Society.

Gerzon, M. (1990). Three channels: the future of stereo? *Studio Sound*, June, pp. 112–125.

Gerzon, M. (1992). Psychoacoustic decoders for multispeaker stereo and surround sound. Presented at *103rd AES Convention, San Francisco, 1–4 October*. Preprint 3406. Audio Engineering Society.

Gerzon, M. and Barton, G. (1992). Ambisonic decoders for HDTV. Presented at *92nd AES Convention, Vienna*. Preprint 3345. Audio Engineering Society.

Holman, T. (1996). Channel crossing. *Studio Sound*, February, pp. 40–42.

ITU-R (1993). *Recommendation BS. 775: Multi-channel stereophonic sound system with or without accompanying picture*. International Telecommunications Union.

Jot, J-M., Larcher, V. and Pernaux, J-M. (1999). A comparative study of 3D audio encoding and rendering techniques. In *Proceedings of the AES 16th International Conference, Rovaniemi, 10–12 April*, pp. 281–300. Audio Engineering Society.

Smyth, S. *et al.* (1996). DTS coherent acoustics: delivering high quality multichannel sound to the consumer. Presented at *100th AES Convention, Copenhagen, 11–14 May*. Workshop 4a-3.

Todd, C. *et al.* (1994) Flexible perceptual coding for audio transmission and storage. Presented at *96th AES Convention*. Preprint 3796.

5 Spatial sound monitoring

The acoustics of monitoring environments such as listening rooms and control rooms, as well as consumer listening situations, is a large subject that could fill a whole book on its own (indeed it has done). In this book discussion is limited to considering the differences between two-channel stereo monitoring and surround sound systems, comparing the various international proposals that have been made for surround sound listening environments in particular. The issue of monitor system alignment is also touched upon, although it is recognised that this subject is the source of some disagreement at the time of writing.

5.1 Introduction to listening room acoustics

5.1.1 Overview

It is generally agreed that while listening rooms should not unduly influence the ability of the sound engineer to monitor a recording, neither should they be completely anechoic (free field rooms with no reflections). Although one might naively think that the best listening environment would be one in which one heard the direct sound from the loudspeakers and nothing else, in reality this is neither desirable nor practical. A number of factors lead to this conclusion. Firstly, although anechoic environments are useful for some laboratory experiments, they do not represent the natural situation in which most people listen to recordings, and a traditional argument goes that it is desirable to monitor the sound in a space that has acoustics not too different from those of the typical listening environment. (Some, though, argue that there is no such thing as a typical domestic environment, and therefore for professional purposes one should ignore it and concentrate on accurate reproduction of what is recorded.) Secondly, sound level falls off rapidly with distance from a loudspeaker in the free field (about 6 dB per doubling in distance), requiring exceptionally high power handling from monitoring systems in order to produce suffi-ciently high sound pressure levels with low distortion at the listening position in an anechoic room. Thirdly, anechoic rooms are exceptionally uncomfortable to work in for any length of time, being tiring and unnatural. Consequently professional sound monitoring environments generally have some reverber-ation and reflections, although these are controlled to be within certain limits. The interested reader is referred to Philip Newell's book *Studio Monitoring Design*, for a substantial coverage of relevant issues, as well as to Alton-Everest's *The Master Handbook of Acoustics*.

5.1.2 Control of reflections

A number of different approaches to the design of monitoring environments have been proposed. Many adhere to the principle that early reflections arriving at the listening position within about 15–20 ms after the direct sound from the loudspeakers are to be minimised in order to avoid spatial imaging and timbral modifications that could otherwise arise from the perceptual interaction of the direct and reflected sound.

The effect of these reflections depends greatly on the nature of the signal, the direction and effect of the reflection, and the task that the listener is asked to perform (there is a difference between asking someone 'can you hear a difference' and asking them to identify a particular effect of the reflection). Owing to the percep-tual integration that takes place between direct sound and reflec-tions, described in Chapter 2, early reflections are rarely heard as discrete echoes but rather as changes in sound quality and spatial image quality. Toole and Olive (1989) investigated the audibility and effect of reflections in rooms and some of their results are summarised in Figure 5.1, together with a summary they published of previous work on detectability of reflections. They compared the effects of a single lateral reflection in three rooms – an anechoic chamber, a 'relatively reflection-free' listening room and an IEC standard listening room. The simulated room reflections clearly had spatial effects above a certain level. It was found that transient signals made it easier to detect reflections than more continuous signals like noise and music. For speech, reflections within the first 20 ms or so were audible once they were above about −15 dB with relation to the direct sound, whereas typical music and noise signals showed detection thresholds at about −20 dB or below. Not surprisingly, the simulated reflection was easiest to detect in the anechoic chamber. Above this threshold the effect was primarily one of timbral change and then increasing spaciousness and phantom image spreading.

Bech (1995) found noticeable changes in timbre resulting from the simulation of reflections in an absorptive space. He also summarised his conclusions regarding the effects of simulated reflections on spatial properties of reproduced sound in small rooms in another paper (Bech, 1998), as:

1. Subjects can reliably discriminate between spatial and timbre cues.
2. The spectral energy above 2 kHz of individual reflections determines the degree of influence the reflection will have on the spatial aspects of the reproduced sound field.
3. Under conditions as in the simulated room, with a standard two-way loudspeaker system reproducing broadband noise or speech, only the first order floor reflection is so strong that it will contribute separately to the spatial aspects of the sound field.

The results of a number of studies suggest that reflections from the sides of rooms are likely to have most effect on the perception of spaciousness and stereo imaging, but such reflections are often too weak in most real control rooms to be above the threshold required to produce a noticeable effect, provided that care is taken with siting of equipment and hard surfaces.

As a result of these investigations, a number of international standards now specify that in rooms designed for critical listening the early reflections arriving at the listening position within the first 15 ms should be at least 10 dB below the direct sound (between 1 and 8 kHz). This is in fact quite hard to engineer in practice, particularly in control rooms with a mixing console between the listener and the loudspeakers, although, to be fair, these standards are not primarily designed for sound mixing rooms but for international standard listening rooms that are normally devoid of anything except the listener and the loudspeakers. Floor and ceiling reflections can be most problematic as they are generally the earliest and the strongest (apart from those resulting from any mixing console surface).

Figure 5.1 Subjective effects of a simulated lateral reflection (65°) on speech reproduced in three different rooms. (RRF listening room is a 'relatively reflection free' room with controlled reflections.) (a) Absolute detection threshold. Above this the reflection appears to cause a change in perceived spaciousness. (b) Image shift threshold. Above this the perceived sound image location begins to be shifted or spread. The reflection level required for these effects clearly depends on the delay. (c) Comparison of absolute detection thresholds found by different authors for various sound sources and reflection angles. (After Toole and Olive, 1989).

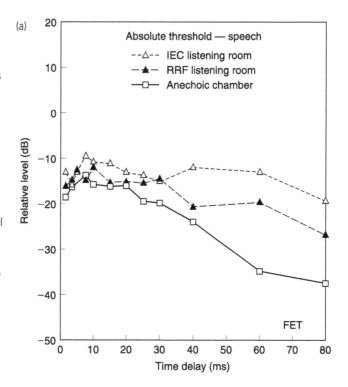

(a)

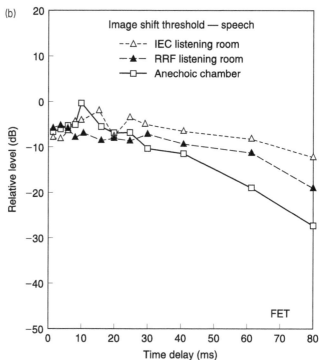

(b)

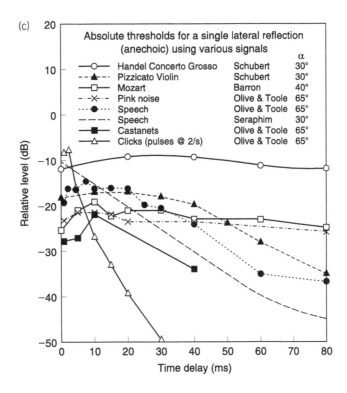

5.1.3 *Low frequency interaction between loudspeakers and rooms*

At low frequencies the response of rooms containing reflections is dominated by room modes. Static patterns of pressure maxima and minima are distributed around the room, resulting from acoustic reflection paths that are related directly to the wavelength of the sound. At low frequencies, loudspeakers in a room are primarily perceived through their coupling with the room modes and the physical position of the speakers with relation to the pressure patterns of the modes governs the degree to which this coupling takes place. If a loudspeaker is placed at a pressure minimum (a node) then it will couple weakly or not at all to the mode, whereas it will couple strongly when placed at a maximum (antinode). This has a substantial effect on the perceived frequency response of the loudspeaker in the room.

Griesinger (1997) has proposed that interference between medial (front–back and up–down) and lateral modes (side–side) will strongly affect the degree of low frequency spaciousness and envelopment perceived in small listening rooms. He claims that if asymmetric lateral modes (those with nulls in the centre of the room) are strong in relation to medial modes the low frequency spaciousness will be high, possibly tending towards phasiness. Asymmetric lateral modes, he asserts, are excited by the antiphase components between loudspeakers, whereas medial modes are excited by the in-phase components of the loudspeaker signals. Medial modes are likely to be suppressed if fronts and or backs of control rooms are made highly absorbent compared to the sides (e.g. using bass traps at the back of the room) or if ceilings are used as bass traps. Vertical modes are also suppressed by the use of dipole (bi-directional) loudspeakers which may give rise to excessive low frequency spaciousness (phasiness) when listening on the centre line of the room.

5.1.4 Two-channel mixing rooms

Two-channel stereo monitoring systems in studios are usually installed either in the boundaries of the room (flush mounted), to either side of the control room window, or free-standing behind the mixing console. Near-field monitors are often used as well, mounted on the mixing console, to give an alternative form of listening that is less affected by the room acoustics and possibly more similar to some forms of domestic listening. Methods of room design became popular in the 1970s and '80s that engineered an early time period after the direct sound from the loudspeakers within which reflections were minimised, in order that early reflections from the room did not unduly modify the perceived sound from the loudspeakers. The so-called LEDE (live-end–dead-end) design was used quite widely, in which the area around the front of the room was treated with absorbent material to minimise early reflections from the front, ceiling and sides, while the rear of the room was more reflective and used diffusers to create later diffuse reverberation around the listening position (see Figure 5.2). This gave rooms a natural sound with controlled early reflections, although reflections off the mixing console surface and control room window are still hard to avoid.

Also experimented with were so-called reflection-free zones, described variously by Bob Walker of the BBC, and Don and Chips Davis in the USA, in which the area around the loudspeakers was reflective, but shaped in such a way as to direct the first few reflections (normally the strongest) away from the listening position and towards the rear of the room, creating a period of time in which the response at the listening position was close to anechoic, as shown in Figure 5.3. Another approach, mentioned

Figure 5.2 Live-end–dead-end (LEDE) principle of control room construction.

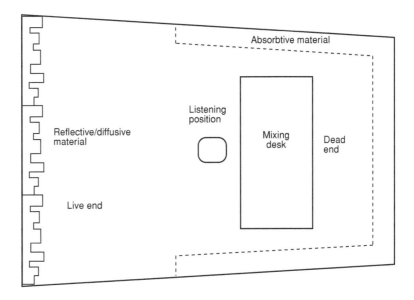

Figure 5.3 Reflection-free zone design according to Bob Walker (BBC) directs early reflections away from listening position.

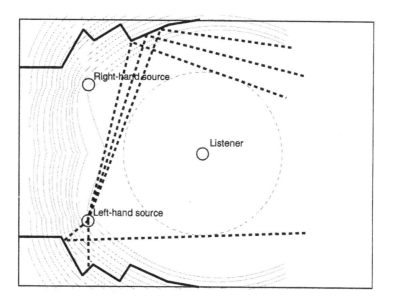

by Varla *et al.* (1999) is to install the loudspeakers flush with a hard front wall, using it as an extended baffle, and making the back wall a thick, soft bass trap (low frequency absorption). This latter approach has been developed quite widely by Tom Hidley and others to create the concept of the 'non-environment' room. Such a room tends towards being anechoic as far as the sound from the front monitors is concerned, since the rest of the room including the ceiling is very heavily treated with absorbent material (the inner shell of the room is spaced a long way from the outer solid shell and the gap is used for a lot of absorbtion. Reflections from the area around the loudspeakers provide some natural sense of space for the people in the room, but this reflective surface is not acoustically 'seen' by the front loudspeakers (which then radiate into an almost anechoic space). Such rooms have been found to result in highly precise imaging of amplitude-panned phantom sources.

As Philip Newell summarises in a recent *Studio Sound* article (Newell, 1998), a large number of rooms designed for two-channel monitoring are essentially what he calls 'bi-directional' – in other words one end is not the same as the other, although lateral symmetry is often maintained. This presents some difficulties when considering the use of such rooms for surround sound monitoring.

5.1.5 Differences between two-channel and multichannel mixing rooms

As there is no universal agreement about how to design two-channel rooms there is even less about the best approach to use for multichannel rooms. While there is a gradual consensus building around the view that rooms for multichannel monitoring should have an even distribution of absorbing and diffusing material so that the rear loudspeakers function in a similar acoustic environment to the front, some people get very hot under the collar about the idea, probably because it suggests that half the music mixing rooms in the world will be unsuitable for surround monitoring! The problem appears to be that it is difficult to make a room that is both optimal for surround *and* optimal for two-channel stereo at the same time.

Other reasons for disagreements about the acoustics of surround control rooms relate to the differences of opinion about the role of the surround channels. If, as the standards suggest, the surround loudspeakers are placed at about ±110°, they become more like side loudspeakers, and reflections off the opposite side walls become more important than those from the front of the room. Alternatively if, as often seems to be the case in some practical rooms, the surround loudspeakers are mounted somewhere near the back corners of the room or behind the mixing position, reflections from the control room window and the front wall will be more of an issue (particularly important if the room is of the type with a reflective area around the front loudspeakers).

Different views about the role of the surround channels in standard 3-2 formats may lead to music balancers attempting to mix discrete, localised sources to the surround channels, treating them as 'equals' to the front channels. In fact, as described earlier in the book, the primary intention for these channels is for non-localisable ambience and effects information that adds to spatial impression. (In the words of the ITU standard: 'it is not required that the side/rear loudspeakers should be capable of prescribed image locations outside the range of the front loudspeakers.') If this concept is followed the issue of a potentially different acoustic environment for the surround loudspeakers is perhaps less of a problem than might at first be thought, since the acoustics affecting the front loudspeakers can be kept roughly as they would be in two-channel monitoring and the rear channels may not suffer too much from the non-ideal reflections they experience. Furthermore, signals fed to the rear channels may typically be at lower levels than those sent to the front, making their reflections correspondingly lower in level, and possibly masked by front channel signals. A lot depends on what one intends to do with the rear channels in mixing technique.

The effects of the acoustics of the control room on surround channels may be ameliorated somewhat if a more distributed array of surround loudspeakers is used, preferably with some form of decorrelation between them to avoid strong comb filtering effects (and appropriate gain/EQ modification to compensate for the summing of their outputs). This is more akin to the film sound situation, though, and may only be possible in larger dubbing stages. In smaller control rooms used for music and broadcast mixing the space may not exist for such arrays. The ITU standard allows for more than one surround loudspeaker on either side and recommends that they are spaced equally on an arc from 60–150° from the front.

One of the difficulties of installing loudspeaker layouts according to the ITU standard, with equal spacing from the listening position and the surrounds at 110° ±10°, is the required width of the space. This arrangement often makes it appropriate for the room to be layed out 'wide' rather than 'long' (as it might be for two-channel setups). If the room is one that was previously designed for two-channel stereo the rotation of the axis of symmetry may result in the acoustic treatment being inappropriately distributed. Also the location of doors and windows may make the modification of existing rooms difficult.

If building a new room for surround monitoring then it is obviously possible to start from scratch and make the room wide enough to accommodate the surround loudspeakers in the right places, and to distribute the acoustic absorption and diffusion more uniformly around the surfaces than might perhaps be the case in two-channel rooms. Varla *et al.* (1999) discuss some of the difficulties inherent in designing such rooms, and this is an interesting review of the effects of reflections from different types of loudspeaker mounting. Bell (2000) also offers some useful practical points from his experience of installing surround sound control rooms.

5.2 International guidelines for surround sound room acoustics

A number of attempts have been made by audio engineering groups to produce guidelines for surround sound listening or monitoring environments. Differences exist between those that are based on existing standards for critical listening environments for programme interchange and quality evaluation work (e.g. listening room specifications) and those that are intended as more down-to-earth practical recommendations for mixing rooms. As a rule it is not possible to conform to all the acoustic specifications of international standard listening rooms in practical mixing spaces that may have equipment in them, although some of the criteria may be met.

The most developed guidelines to date have come from the German Surround Sound Forum (whose proposals are based on international reference listening room standards), the Japanese HDTV Forum (whose proposals are based on the need for more practical mixing room guidelines), and the AES Technical Committee on Multichannel and Binaural Audio of which the author is chairman (which is attempting to summarise examples of good practice from a number of proposals). THX also has comprehensive acoustic guidelines for rooms that are designed to conform to its proprietary recommendations. It is unclear at the present time whether recommendations for film mixing rooms and smaller music or broadcast mixing rooms can be harmonised. The existence of standards or guidelines for room acoustics by no means guarantees that they will sound the same. Indeed the author's experience of three international standard listening rooms conforming to the most stringent ITU standard (BS.1116) is that there is enough leeway in the standard for them to sound noticeably different to each other. Nonetheless such guidelines do limit the options and provide ranges for reverberation time, noise levels, reflection levels and dimension ratios.

5.2.1 Suggestions for listening room acoustics based on international standards

The tables in this section are based on the work carried out by the German Surround Sound Forum and proposed for inclusion in the AES technical committee document. They relate to reference listening conditions, providing a basis for international comparison of sound programme material, and may be difficult to meet in some practical mixing rooms. Mostly they are based on existing ITU and EBU standards (EBU Tech 3276-E, including Supplement 1 for multichannel systems, is specified for 'the assessment of sound programme material').

Table 5.1 shows the dimensions suggested for reference listening rooms. A volume of 300 m³ should not be exceeded. The dimension ratios suggested are designed to create a suitable distribution of room modes. The room is typically symmetrical

Table 5.1 Typical requirements for a reference listening room

Parameter	Units/conditions	Reference listening conditions
Room size		
Floor surface area:		
Mono/2-channel-stereo	S [m²]	>30
Multichannel		>40
Room proportions	l = Length	1.1w/h ≤ l/h ≤ 4.5w/h-4
	w = Width	with
	h = Height	l/h<3 and w/h<3
		avoiding dimension ratios that are
		within 5% of integer values
Base width		
2-channel stereo	B [m]	2.0–4.0
Multichannel		2.0–4.0
Basis angle		
2-channel-stereo	[degrees]	60
Multichannel	referred to L/R	60
Listening distance		
2-channel stereo	D [m] from	Between 2m and 1.7 times B
Multichannel	acoustic centre	
Listening zone		
2-channel stereo	R (radius) [m]	0.8
Multichannel		0.8
Loudspeaker height		
2-channel stereo	h [m]	≈ 1.2
Multichannel (all)		≈ 1.2
Distance to surrounding reflecting surfaces		
2-channel stereo	d [m]	≥ 1
Multichannel		≥ 1

Table 5.2 Reference sound field conditions

Parameter	Units/conditions	Reference listening room
Direct sound		
Amplitude/frequency response	Free field propagation measurements	Tolerance borders see Table 5.3 (reference monitor)
Reflected sound		
Early reflections	0–15 ms (in the region 1 kHz to 8 kHz)	<–10 dB relative to direct sound
Temporary diffusion of the reverberant sound field	Avoidance of significant anomalies in the sound field.	No flutter echoes, no sound colouration etc.
Reverberation time	T_m (s) – Nominal value in the region of 200 Hz to 4 kHz. V = Volume of listening room; V_0 = Reference room volume of 100 m³.	$\approx 0.25^* (V/V_0)^{1/3}$ (For tolerance range see Figure 5.4.)
Stationary sound field		
Operational sound level curve level curve	50 Hz–2 kHz 2 kHz–16 kHz	±3 dB ±3 dB from –3 dB to –6 dB in accordance with tolerance shown in Figure 5.5.
Background noise		Ideally <NR 10 but not exceeding NR15
Reference listening level (relative to defined measurement signal)	Input signal: Pink Noise, –18 dBFS (RMS)	78 dBA (RMS slow) (per channel)

around the listening direction, and should take into account the distribution of absorption material, especially around the speakers, doors, windows, and technical equipment, etc., so that any acoustical discontinuities can be avoided. The surface of any mixing desk should be designed to avoid disturbing reflections, as far as possible.

Table 5.2 shows the sound field conditions that are desirable at the listening position. Figure 5.4 shows the limits within which the reverberation time of the room should ideally be held. Notice that it allows for considerably longer RT at low frequencies where it is often difficult to get sufficient absorption. The measurements are made using the loudspeakers in the room and with 1/3-octave band filtering. T_m is the average of the measured reverberation time T in the 1/3-octave bands from 200 Hz to 4 kHz. It should lie between 0.2 and 0.4 seconds, depending on the room size. According to the standards from which these recommendations are taken, the frequency response for the reverberation time should be steady and continuous; sudden or strong breaks should be avoided. Therefore deviations in adjoin-

Figure 5.4 Suggested reverberation time limits for surround sound listening rooms (AES, after German Surround Sound Forum). T_m is the average value between 200 Hz and 4 kHz and should ideally lie between 0.2 and 0.4 seconds. The heavy black lines indicate the suggested limits and the bold dotted lines indicate suggested tighter tolerances at the extremes.

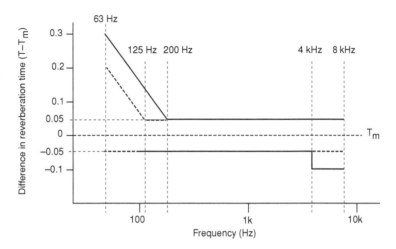

ing 1/3-octave bands in the region of 200 Hz to 8 kHz should not exceed 0.05 s, and under 200 Hz less than 25% of the longest reverb time should not be exceeded.

An operational sound level curve is also defined (see Figure 5.5). This shows the frequency response of the monitor loudspeakers at the listening position, including any effects of the room, and may require equalisation if it is to be achieved. It is defined as the frequency response of the sound pressure level at the reference listening position, using band filtered pink noise for each loudspeaker separately.

The background noise level (from air conditioning or other external or internal sound sources) is given in form of 1/3-octave band sound pressure level $L_{pFeq,\ T\text{-}30s}$ (RMS, slow), in accordance

Figure 5.5 Suggested operational room response tolerances of loudspeakers at listening position (AES, after German Surround Sound Forum). L_m is the mean value.

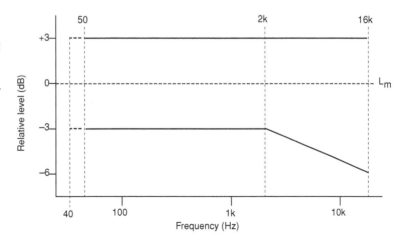

Table 5.3 Typical requirements for reference monitor loudspeakers

Parameters	Units/conditions	Smaller room	Larger room
Amplitude/frequency response	40 Hz ...16 kHz		
response	0°	Tolerance: 4 dB	Tolerance: 4 dB
	±10°	Deviation to 0°: 3 dB	Deviation to 0°: 3 dB
	horizontal ±0°	Deviation to 0°: 4 dB	Deviation to 0°: 4 dB
Difference between front loudspeakers	in the range > 250 Hz to 2 kHz	0.5 dB	0.5 dB
Directivity index	250 Hz ... 16 kHz	8 dB ± 2 dB	6–12 dB
Non-linear distortion	<100 Hz	−30 dB (= 3%)	−30 dB (= 3%)
attenuation (SPL = 96 dB)	>100 Hz	−40 dB (= 1%)	−40 dB (= 1%)
Transient fidelity			
Decay time t_s, for reduction to a level of 1/e, i.e. 0.37 of the output level	t_s [s]	<5/f [Hz] (preferable: 2.5/f)	<5/f [Hz] (preferable: 2.5/f)
Time delay			
Difference between loudspeakers	δt	≤ 10 µs	≤ 10 µs
Dynamic range			
Maximum operating level (measurement acc. to IEC 268-5, § 17.2, referred to 1 meter distance)	$L_{eff\ max}$	> 112 dB (using IEC 268-1 programme simulation noise or special condition)	> 120 dB (using IEC 268-1 programme simulation noise or special condition)
Noise level	L_{noise}	≤ 10 dBA	≤ 10 dBA

with ISO noise rating (NR) curves for the 1/3-octave band averaged frequencies from 50 Hz to 10 kHz. These guidelines suggest that NR10 is desirable and that NR15 should not be exceeded.

The specifications in Table 5.3 relate to reference monitor loudspeakers. The German Forum points out that there are loudspeakers that comply with these requirements that are not necessarily suitable for all programme genres as reference loudspeakers, and that the conclusive selection and decision is formed on the strength of investigative subjective tests and the resulting criteria and attributes.

The *amplitude/frequency response* is measured under free field conditions with pink noise for the 1/3-octave band averaged frequencies in the range 31.5 Hz to 16 kHz at 0º, ±10º and ±30º. It is recommended that the response is symmetrical around the reference axis.

The *directivity index* can also be derived from the 1/3-octave band measurements. It can either be calculated from the direc-

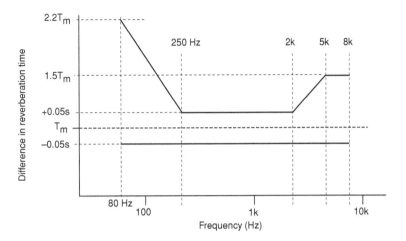

Figure 5.6 Suggested reverberation time limits for surround sound mixing rooms (AES, after Japanese HDTV forum). T_m is the average value between 250 Hz and 2 kHz.

tional characteristics or derived from the difference between the free field measurements and the diffuse field measurements. The ITU indicates that a directivity index of >6 dB with a steady slow increase towards higher frequencies is desirable. The standards currently recommend using identical loudspeakers for all the five channels, for compatibility purposes.

5.2.2 Suggestions for mixing room acoustics

The Japanese HDTV forum has developed guidelines for practical rooms designed for mixing multichannel audio for high

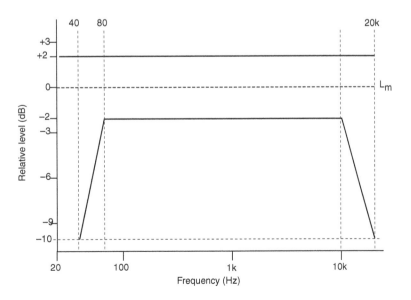

Figure 5.7 Suggested anechoic monitor response tolerances (AES, after Japanese HDTV forum).

Table 5.4 Multichannel mixing room specifications according to Japanese HDTV surround forum

Parameters		Designing guideline	
		Small room	Medium room
Display		CRT – 36 inches	Acoustically transparent (perforated) screen – 140 inches
Room	Floor area m²	50±20	100±30
	Room volume m³	Equal to or more than 80	Equal to or more than 200
	Room shape	Non-rectangular (avoid parallel surfaces)	
	Dimensional ratios	Avoid ratios with simple integral numbers. H:W:L = 1:1.59±0.7:2.52±0.28. etc. are desirable	
	Room height (m)	3.0~4.0	4.0~6.0
Interior finish		Uniform absorbent/diffusively reflective treatment to avoid strong reflections from specific directions	
Acoustical properties	Reverberation time (s)	0.2±0.05 at 500 Hz	0.3±0.1 at 500 Hz
	Mean absorption coefficient	0.4~0.6 at 500 Hz	
	Reverberation characteristics	See Figure 5.6	
	Static transfer frequency response	±3 dB (one octave band) between 125 Hz and 4 kHz. Up to 2 bands may be within ±4 dB	
	Early reflections	Any reflections within 15 ms after the direct sound should be lower by 10 dB relative to the direct sound	
	Interaural cross-correlation	Not specified (under consideration)	
	Distribution of the SPL	Uniform SPL within the listening area including the mixing point	
Noise	Air conditioning noise	Noise Criterion Curve of NC-15 (NR-15 would be desirable)	
	Equipment/background noise	Noise Criterion Curve of NC-20 (NR-20 would be desirable) The fan noise of video projector etc. should be reduced	

Loudspeaker arrangement

L/R	Setting	Flush mounting is desirable to avoid reflections from surrounding walls etc. Eliminate these reflections in case of free standing	
	Axis direction (Reference Point)	Mixing position or zero to one metre behind it	
	Distance(L~R) m	3.0~6.0	5.0~8.0
	Height (m)*	1.2~2.0*²	Centre of the screen*²
	Distance to the reference point	All the distances from L/C/R/SL/SR loudspeakers to the reference point would be desired to be equal	
	Subtended angle against the centre line of the room (degrees)	30	30
C	Setting	Flush mounting is desirable to avoid reflections from surrounding walls etc. Eliminate these reflections in case of free standing	
	Axis direction (Reference Point)	Mixing position or zero to 1 metre behind it	
	Height (m)*¹	The same height as the L/R is desirable*³	Centre of the screen*³
	Distance to the reference point	All the distances from L/C/R/SL/SR loudspeakers to the reference point should be equal	
S_L/S_R	Number	Equal to or more than two	Equal to or more than four
	Setting	Flush mounting is desirable but being attached to the wall is acceptable because of the room shape etc.	
	Axis direction (Reference Point)	Mixing position or zero to 1 metre behind it	
	Height (m)*¹	Same or higher than the L/R is desirable L/R*(0.9~1.4)*⁴	

continued

	Parameters	Designing guideline	
		Small room	Medium room
	Distance to the reference point	All the distances from L/C/R/S$_L$/S$_R$ loudspeakers to the reference point should be equal	
S	Subtended angle against the centre line of the room (degrees)	120±10	More than 110 (symmetrically dispersed at regular intervals)
Monitoring level		85±2 dB (C weighted)/ch (pink noise) at −18 dB dBFS for large loudspeaker 80±2 dB (C weighted)/ch (pink noise) at −18 dB dBFS for medium loudspeaker 78±2 dB (C weighted)/ch (pink noise) at −18 dBFS for small loudspeaker	
Monitor loudspeaker			
Maximum sound pressure level[5]	L/C/R	Equal to or more than 117 dB	Equal to or more than 120 dB
	When 2 surround speakers	Equal to or more than 114 dB	Equal to or more than 117 dB
	When 4 surround speakers	Equal to or more than 111 dB	Equal to or more than 114 dB
	When 8 surround speakers	Equal to or more than 108 dB	Equal to or more than 111 dB
Amplitude versus frequency response	L/C/R	See Figure 5.7	
Effective frequency range[6]	L/C/R	40 Hz~20 kHz	
	SL/SR	Same as L/C/R, at least between 80 Hz~20 kHz	
Non-linear distortion[7]	L/C/R	<3% for 40 Hz~250 Hz, <1% for 250 Hz~16 kHz	
	SL/SR	Same as L/C/R At least <3% for 80 Hz~250 Hz, <1% for 250 Hz~16 kHz	
Transient fidelity	L/C/R/S$_L$/S$_R$	The decay time to the level of 1/e (approximately 0.37) from the original level should be less than 5/f (where f is frequency)	
Phase, Group delay[8,9]	L/C/R/S$_L$/S$_R$	Either of them would desirably be indicated	
Directivity index	L/C/R/S$_L$/S$_R$	6–12 dB (ITU-R BS.1116-1)	
Impedance	L/C/R/S$_L$/S$_R$	>3.2Ω	
Deviation of frequency response	L/C/R/S$_L$/S$_R$	<1.5 dB for 100 Hz~10 kHz Peak/dip narrower than 1/3 oct. shall be neglected.	
Efficiency[10]	L/C/R/S$_L$/S$_R$	Should be indicated	

Notes:

*1 Height of loudspeakers: Height of acoustical centre of the loudspeaker from the floor level at mixing position.

*2 More than 1.2 metres is recommended. But the height may be 1.7 metres to avoid the metre bridge of high console shadowing the direct sound and that may be 1.9 metres when the loudspeakers are set above the window.

*3 When the C loudspeaker is set below the CRT, its height may be lower than the L/R loudspeakers.

*4 Same as L/C/R is desirable, but it could be 2.2~2.7 metres because of doors on the side or rear walls.

*5 Maximum sound pressure level = (Rated output sound pressure level) + (Maximum input level).

*6 Effective frequency range (between −10 dB points).

*7 Absolute sound level is measured at 1 m from the loudspeaker.

*8 Directivity index of the front loudspeakers depends on the programme.

*9 Difference of overall impressions caused by the directivity index of the rear loudspeakers is rather small.

*10 Efficiency is indicated by the rated output sound pressure level at (1m, 1w).

definition television applications, but the criteria are also useful for other multichannel mixing applications. These proposals are slightly looser in one or two areas than those specified above for reference listening rooms, and may be more practical for use in sound mixing rooms with equipment present.

Table 5.4 summarises the Japanese recommendations, split into small and medium rooms. Figures 5.6 and 5.7 show the suggestions for reverberation time and anechoic monitor frequency response. (Note the important difference between anechoic monitor response and the operational room response that was shown in Figure 5.5. Operational room response includes the effects of the listening room.) The noise levels allowed are slightly higher than those shown for reference listening rooms. The left–right loudspeaker spacing is also considerably greater (possibly too large for some applications) and there is more leeway in the height of the loudspeakers, particularly in the case of LS and RS channels where doors and other room features may prevent the speakers being at the same height as the front speakers.

5.2.3 Proprietary certification standards

Commercial certification of surround mixing rooms has become quite popular in recent years. THX, the company that developed proprietary acoustic and equipment standards for cinemas and sound mixing stages for film, has developed a specification known as 'PM3' for 'professional multichannel mixing and monitoring'. The company works with studios to ensure certain standards of acoustics, monitoring and amplification for surround installations, for sound-only or sound-with-picture applications. Some parts of the PM3 certification are more flexible than others, and the company certifies both small mix rooms and larger dubbing stages, taking into account the need to undertake 'absolute' or 'relative' judgement of sound quality. A proprietary crossover controller unit known as the CC4 deals with bass management and also with switching of monitor/mixdown mode between formats such as 5.1, 7.1 and conventional stereo.

5.3 Loudspeakers for surround sound: placement and directivity

There is some debate over the type and placement of loudspeakers for surround sound monitoring purposes. This is partly due to practical problems encountered when trying to install multichannel monitoring in two-channel rooms, and partly because of debates about the directivity characteristics required of surround loudspeakers. There are also substantial differences between film sound mixing in large spaces and mixing for small rooms (e.g. music or television).

5.3.1 Flush-mounted versus free-standing loudspeakers

In many studios it is traditional to mount the monitor loudspeakers flush with the front wall. This has the particular advantage of avoiding the reflection that occurs with free-standing loudspeakers from the wall behind the loudspeaker, causing a degree of cancellation at a frequency where the spacing is equal to one quarter of the radiated wavelength. It also improves the low frequency radiation conditions if the front walls are hard. Nonetheless, it is hard to find places to mount five large loudspeakers in a flush-mounted configuration, and such mounting methods can be expensive. Furthermore the problems noted above, of detrimental reflections from rear loudspeakers off a hard front wall or speaker enclosure, can arise, depending on the angle of the rear loudspeakers. For such reasons, some sources recommend making the surfaces around the loudspeakers reflective at low frequencies and absorbent and mid and high frequencies.

The problem of low frequency cancellation notches with free-standing loudspeakers can be alleviated but not completely removed. The perceived depth of the notch depends on the absorption of the surface and the directivity of the loudspeaker. By adjusting the spacing between the speaker and the wall, the frequency of the notch can be moved (downwards by making the distance greater), but the distance needed is often too great to be practical. If the speaker is moved close to the wall the notch position rises in frequency. This can be satisfactory for large loudspeakers whose directivity is high enough at middle frequencies to avoid too much rear radiation, but is a problem for smaller loudspeakers.

The use of a 5.1-channel monitoring arrangement (rather than five full-bandwidth loudspeakers), with proper bass management and crossovers, can in fact ameliorate the problems of free-standing loudspeakers considerably. This is because a subwoofer can be used to handle frequencies below 80–120 Hz and it can be placed in the corner or near a wall where the cancellation problem is minimised (see below). Furthermore, the low frequency range of the main loudspeakers can then be limited so that the cancellation notch mentioned above occurs below their cut-off frequency.

5.3.2 Front loudspeakers in general

As a rule, front loudspeakers can be similar to those used for two-channel stereo, although noting the particular problems with the centre loudspeaker described in the next section. The guidelines above suggest that the directivity index of the front loudspeakers in small rooms should preferably lie between 6 and 10 dB from 250 Hz to 16 kHz. Nonetheless, Munro (1999) has suggested that low directivity for the front loudspeakers may be desirable when trying to emulate the effect of a film mixing situation in a smaller surround control room. This is because in large rooms the sound balancer is often well beyond the critical distance where direct and reflected sound are equal in level, and using speakers with low directivity helps to emulate this scenario in smaller rooms. Film mixers generally want to hear what the large auditorium audience member would hear, and this means being further from the loudspeakers than for small room domestic listening or conventional music mixing.

5.3.3 What to do with the centre loudspeaker

One of the main problems encountered with surround monitoring is that of where to put the centre loudspeaker in a mixing room. Ideally it should be of the same type or quality as the rest of the channels and this can make such speakers quite large. In 5.1 surround setups there is an increasing tendency to use somewhat smaller monitors for the five main channels than would be used for two-channel setups, handling the low bass by means of a subwoofer or two. This makes it more practical to mount a centre loudspeaker behind the mixing console, but its height will often be dictated by a control room window or video monitor (see below). The centre loudspeaker should be on the same arc as that bounding the other loudspeaker positions, as shown in the ITU layout in Chapter 4, otherwise the time delay of its direct sound at the listening position will be different to that of the other channels. If the centre speaker is closer than the left or right channels then it should be delayed slightly to put it back in the correct place acoustically.

In the case of near field arrangements the centre loudspeaker can sometimes sit on the meter bridge of the mixer, but this can require a delay to make it acoustically the same distance from the listener, and it encourages mixer reflections, so spacing it slightly back from the mixer on a stand is ideal. Sometimes such centre speakers are designed slightly differently to left and right speakers, either being orientated on their sides for convenience or being symmetrical about the vertical centre line (some speakers have offset tweeters, for example).

Dolby suggests that bass management in surround decoding can be used to split the low frequency content from the centre channel below, say, 100 Hz, feeding it equally to larger left and right loudspeakers, making it practical to use a smaller unit for the centre loudspeaker, preferably with the same mid and high frequency drivers as the main speakers.

The biggest problem with the centre loudspeaker arises when there is a video display present. A lot of 5.1 surround work is carried out in conjunction with pictures and clearly the display is likely to be in exactly the same place as one wants to put the centre speaker. In cinemas this is normally solved by making the screen acoustically 'transparent' and using front projection, although this transparency is never complete and usually requires some equalisation. In smaller mixing rooms the display is often a flat-screen plasma monitor or a CRT display and these do not allow the same arrangement. Most of the international standards that discuss this problem do not come up with any optimal solutions to this problem, acknowledging that nearly everything one does will be a compromise. For example, the EBU in Tech. 3276, Supplement 1, says:

> The presence of the viewing screen also causes difficulties for the location of the centre loudspeaker. The height of the screen almost always makes it impossible to meet the height and inclination requirements for the loudspeaker. Screens which are 'acoustically transparent' would allow the loudspeaker to be placed in the correct location behind the screen. However, such screens generally cause some alteration of the sound quality, both by attenuation of the direct sound and by causing reflections and standing waves in the space between the rear face of the screen and the front face of the loudspeaker. Sometimes two centre loudspeakers are used, driven in phase, with one above and one below the screen. This arrangement can cause severe response irregularities for listening positions that are not on the horizontal axis of symmetry.

With modestly sized solid displays for television purposes it can be possible to put the centre loudspeaker underneath the display, with the display raised slightly, or above the display angled down slightly. The presence of a mixing console may dictate which of these is possible, and care should be taken to avoid strong reflections from the centre loudspeaker off the console surface. Neither position is ideal and the problem may

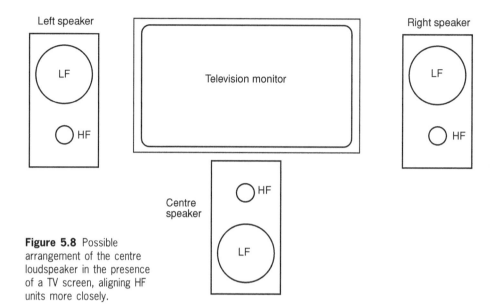

Left speaker

Right speaker

Television monitor

LF

HF

LF

HF

Centre
speaker

HF

LF

Figure 5.8 Possible
arrangement of the centre
loudspeaker in the presence
of a TV screen, aligning HF
units more closely.

not be solved easily. Dolby suggests that if the centre loudspeaker has to be offset height-wise it could be turned upside down compared with the left and right channels to make the tweeters line up, as shown in Figure 5.8. Some vertical misalignment of the centre speaker position is probably acceptable from a perceptual standpoint, as the resolution of the hearing mechanism in this plane allows for a few degrees of difference.

Interestingly, the flat-panel loudspeaker company, NXT, has shown large flat-panel loudspeakers that can double as projection display screens, which may be one way forward if the sound quality of the flat panels speakers can be made high enough.

5.3.4 Surround loudspeakers

Nearly all the standard recommendations for professional setups suggest that the surround loudspeakers should be of the same quality as the front ones. This is partly to ensure a degree of inter-system compatibility. In consumer environments this can be difficult to achieve, and the systems sold at the lower end of the market often incorporate much smaller surround loudspeakers than front. As mentioned above, the use of a separate loudspeaker to handle the low bass (a so-called 'subwoofer') may help to ameliorate this situation, as it makes the required volume of all the main speakers quite a lot smaller. Indeed Bose has had considerable success with a consumer system involving extremely small satellite speakers for the mid–high frequency content of the replay system, mountable virtually anywhere in the room, coupled with a low frequency driver that can be situated somewhere unobtrusive.

The directivity requirements of the surround loudspeakers have been the basis of some considerable disagreement in recent years. Interested readers are referred to a frank exchange published in the *AES Journal* (Holman and Zacharov, 2000) that debates the subjective and objective performance of 'direct radiator' and 'dipole' loudspeakers in surround setups. The debate centres around the use of the surround loudspeakers to create a diffuse, enveloping sound field – a criterion that tends to favour either decorrelated arrays of direct radiators (speakers that produce their maximum output in the direction of the listener) or dipole surrounds (bi-directional speakers that are typically oriented so that their main axis does not point towards the listener). If the creation of a diffuse, enveloping rear and side sound field is the only role for surround loudspeakers then dipoles can be quite suitable if only two loudspeaker positions are available. If, on the other hand, attempts are to be made at all round source localisation (which, despite the evidence in some literature, is not entirely out of the question), direct radiators are considered more suitable. Given the physical restrictions in the majority of control rooms it is likely that conventional loudspeakers will be more practical to install than dipoles (for the reason that dipoles, by their nature, need to be free-standing, away from the walls) whereas conventional speakers can be mounted flush with surfaces.

A lot depends on the application, since film sound mixing has somewhat different requirements to some other forms of mixing, and is intended for large auditoria. Much music and television sound is intended for small-room listening and is mixed in small rooms. Munro suggests that for smaller rooms designed to emulate film sound mixing environments one can use surround speakers that are identical to the front three, provided they are of low directivity and are aimed so as to avoid directly focused sound at the mix position. He proposes that this is better achieved with a number of speakers that can be smaller and easier to install. This was also the primary motivation behind the use of dipoles in consumer environments – that is the translation of the large-room listening experience into the small room. In large rooms the listener is typically further into the diffuse field than in small rooms, so film mixes made in large dubbing stages tend not to sound right in smaller rooms with highly directional loudspeakers. Dipoles or arrays can help to translate the listening experience of large room mixes into smaller rooms.

5.3.5 Subwoofers

The issues of low frequency interaction between loudspeakers and rooms, mentioned earlier, have a substantial bearing on the placement of subwoofers or low frequency loudspeakers in listening rooms. There appears to be little agreement about the optimum location for a single subwoofer in a listening room, although Nousaine (1997) has shown measurements that suggest a corner location for a single subwoofer provides the most extended, smoothest low frequency response. In choosing the optimum locations for subwoofers one must remember the basic principle that loudspeakers placed in corners tend to give rise to a noticeable bass boost, and couple well to most room modes (because they have antinodes in the corners). Some subwoofers are designed specifically for placement in particular locations whereas others need to be moved around until the most subjectively satisfactory result is obtained. Some artificial equalisation may be required to obtain a reasonably flat overall frequency response at the listening position. Phase shifts or time delay controls are sometimes provided to enable some correction of the time relationship of the subwoofer to other loudspeakers, but this will necessarily be a compromise with a single unit. A subwoofer phase shift can be used to optimise the sum of the subwoofer and main loudspeakers in the crossover region for a flat response.

Zacharov, Bech and Meares (1998) found substantial measured differences between subwoofer positions, in terms of frequency response, but were unable to detect the differences subjectively when listening to a range of multichannel programme material with subwoofers in different positions. Kügler and Theile (1992) compared the use of a single subwoofer in different positions with stereo subwoofers placed under the main two-channel loudspeakers, and found that the detectability of a difference varied with programme material, location and crossover frequency, being most noticeable once the crossover frequency rose much above 120 Hz. Bell (2000) suggests that a single subwoofer should always be placed in the centre, as experience shows that it is easy to locate non-central low frequency images at the subwoofer position which is distracting. The reasons for this can be various, and others have shown that port noise, distortion and information above 120 Hz radiating from the subwoofer position can make it localisable, whereas otherwise it would not be. A centrally located subwoofer, though, is likely to suffer from being at the null of lateral standing wave modes. An offset might therefore be considered acoustically desirable. Bell proposes that if two subwoofers are used they should be symmetrically placed, and that considerable success has been had by distributing some of the <120 Hz information to the main loudspeaker array.

There is some evidence to suggest that multiple low frequency drivers generating decorrelated signals from the original recording create a more natural spatial reproduction than monaural low frequency reproduction from a single driver. Griesinger proposes that if monaural LF content is reproduced it is better done through two units placed to the sides of the listener, driven 90° out of phase, to excite the asymmetrical lateral modes more successfully and improve LF spaciousness.

Others warn of the dangers of multiple low frequency drivers, particularly the problem of mutual coupling between loudspeakers that takes place when the driver spacing is less than about half a wavelength. In such situations the outputs of the drivers couple to produce a level greater than would be predicted from simple summation of the powers. This is due to the way in which the drivers couple to the impedance of the air and the effect that one unit has on the radiation impedance of the other. The effect of this coupling will depend on the positions to which sources are panned between drivers, as described by Newell (2000), affecting the compatibility between the equalisation of mixes made for different numbers of loudspeakers.

5.4 Monitor level alignment

5.4.1 Main channel alignment

Practices differ with regard to the alignment of listening level for two-channel and multichannel reproduction. One might reasonably ask why any 'standard' needs to be specified for the listening level of such material, the only thing of any real importance being the *relative* level between the channels. Absolute listening levels are regarded as important for some applications as they enable sound balancing and quality control to be undertaken with relation to a known reference point. In the film industry this is regarded as important because film theatres are aligned to a universal standard that ensures a certain sound pressure level for a certain recorded level on the film sound track. In this way the relative loudnesses of programmes can be compared to some degree. The same is true for critical listening tests and for broadcast quality control, since small level differences between material can give rise to noticeable changes in timbre and spatial sound quality. The concept of a reference monitor level alignment has not caught on so much in music mixing, where the trend seems to be to listen as loud as one likes for pop mixing and at an arbitrary lower level for classical mixing.

Before going on to discuss methods involving measurement, it should be mentioned that subjective alignment of the relative loudness levels between channels can be quite successful, provided a suitable noise test signal is available that can be routed to each channel in turn. The reason for SPL metering is primarily to set standard listening levels for quality control and comparability purposes.

ITU and EBU standards all tend to work to the same formula for aligning the reference level for critical listening, programme interchange and programme comparison. This relies on the peak recording level of the programme being controlled to within certain limits compared with full modulation of the recording medium (0 dBFS in the digital domain) as described in Section 4.3.6.

It is common to use a pink noise signal (that is a random noise signal with a flat spectrum when measured on a log frequency scale, or one which has equal energy per octave across the audio range) recorded at –18 dBFS RMS (18 dB below digital peak level) for this type of level alignment. In some recommendations the noise signal is filtered to exclude low frequency energy, as discussed below, but the ITU and EBU standards assume broadband noise. According to these standards the level of each reproduction channel individually (excluding the LFE channel) is set so that the sound level (RMS slow) at the reference listening position is:

$$L_{LISTref} = 85 - 10 \log n \text{ (dBA)}$$

Where n is the number of reproduction channels in the relevant configuration. So if one channel has a reference listening level **$L_{LISTref}$ = 78 dBA**, then the five combined channels of the 3/2 multichannel stereo configuration have a resulting reference listening level of **$L_{LISTref}$ = 85 dBA**. In order to check this total level a source of non-coherent pink noise (that is noise which has a random phase relationship between the channels) can be played through all loudspeakers at once. Coherent (in-phase) noise is not recommended because the summing effects of an essentially monophonic signal from multiple loudspeakers at the listening position can result in highly position-dependent peaks and dips in the sound pressure, and this is not representative of normal music signals. One can make a non-coherent multichannel noise tape reasonably easily by recording a number of tracks of pink noise from a single noise generator, each on a separate pass of the tape to ensure that the tracks have a random relationship. Any level difference between channels should not exceed 1 dB and ±0.25 dB is recommended.

An alternative 'film-style' recommendation to the above uses pink noise band-limited between 500 Hz and 2 kHz, at the SMPTE standard alignment level of –20 dBFS. This signal is aligned for an SPL of 83 dBC (slow) at the monitoring position, when setting the level of each channel individually (NB: a –18 dBFS test signal would then read 85 dBC). This ends up slightly louder overall than the ITU-style alignment level mentioned above. In movie theatres and film dubbing stages it is common practice to align the surround channels with a –3 dB offset in gain with respect to the front channels. The recording levels of stereo surround channels are correspondingly increased, as noted in Chapter 4. The Japanese HDTV mix room recommendation described in Section 5.2.2 appears to use broadband pink noise with C-weighted measurement, giving different SPL recommendations depending on the size of loudspeaker in use. The film-style methods of alignment are unlikely to result in the same loudness at the listening position as the ITU/EBU method, but in each case (with the exception of film theatres, where there is the offset of the rear channels) all channels are aligned individually, using a noise signal, for equal weighted SPL at the listening position.

In its Dolby Surround mixing manual, Dolby recommends lowering the monitoring line-up level from 85 dBC to 79 dBC when mixing surround material with dialogue for consumer environments. This apparently encourages mixers to increase the dialogue level in the mix and make it more suitable for home environments where distracting noise levels are higher. The author's experience is also that the alignment levels proposed in all these various standards result in programme loudness that is often judged to be excessively high by consumer listeners. To emulate more realistic consumer hi-fi listening levels one may need to align systems for reference level SPLs between 68–75 dB, depending on peak recording level.

The ideal bandwidth of the noise signal used in alignment is the subject of some debate. As exemplified above, some standard recommendations for level alignment recommend the use of broad-band pink noise, or pink noise band limited from

Table 5.5 A selection of different test signals and respective measurement weighting filters

Test signal	Pink noise filtering	Measuring method
Dolby AC-3	500–1000 Hz, 9 dB per octave outside this range	dBC
TMH Labs	500–2000 Hz, 18 dB per octave outside range	dBC
ITU BS.1116	20 Hz–20 kHz	dBA
German SSF	200 Hz–20 kHz	dBA
Japanese HDTV	20 Hz–20 kHz	dBC

200 Hz–20 kHz. This has been criticised by some for involving too much low frequency content, and thereby making the measurement strongly dependent on room mode response, as well as being very direction-dependent at HF. Some low frequency roll-off is often considered to be desirable but the precise frequency of this roll-off is not agreed. Furthermore, while some proponents have also recommended band-limiting at HF as well (e.g. 2 or 4 kHz) others have proposed no HF limit (noise extending to 20 kHz). Some of the alternatives are shown in Table 5.5. Those standards recommending non-band-limited noise signals normally measure the SPL with an A-weighting filter which reduces the extreme LF and HF components considerably, while those recommending band-limited signals often use C-weighting. (C-weighting is a somewhat 'flatter' curve that approximates the equal loudness contours at higher levels than A-weighting, and seems to be more popular in the film sound domain.)

Research conducted during the EUREKA 1653 (MEDUSA) project, by Bech, Zacharov and others (e.g. Zacharov and Bech, 2000), attempted to find correlations between subjective alignment of channel loudness and a variety of objective measurements, using a wide range of different test signals. Although some earlier work had indicated that B-weighted pink noise measurements and subjective adjustments of channel loudness might be quite closely related, recent experiments appear to show less clear distinction between the various measuring methods or test signals. There is some evidence, though, that the low frequency content of a test signal is ignored by listeners when subjectively aligning channel gain.

Dolby has developed a meter for measuring the loudness of programme material based on a modified CCIR-468 noise weighting curve that they find corresponds reasonably well to perceived loudness of programmes.

5.4.2 LFE channel level alignment

The LFE channel of a 5.1 surround system is designed to be aligned so that its in-band gain on reproduction is 10 dB higher than that of the other channels. This does not mean that the overall subwoofer output should have its level raised by 10 dB compared with the other channels, as this would incorrectly boost any LF information routed to the subwoofer from the main channels. Pink noise sent to or recorded on the LFE channel, filtered to the bandwidth of the LFE channel (usually 120 Hz), should be aligned so that its reproduced SPL in individual one-third octave bands over this range is 10 dB above that of the other channels, for a given recorded signal level.

5.4.3 Monitor equalisation

Various acousticians and recording engineers disagree about the use of monitor equalisation. If it is used it is used to create a particular frequency response curve at the listening position, and preferably some places around that. Some prefer not to use such equalisation as it can introduce additional phase anomalies and more equipment into the monitor signal chain, but modern digital equalisers make it more practical. Traditionally it has been done using pink noise, a measuring microphone and a spectrum analyser, with a one-third octave graphic equaliser to adjust the frequency response. For simple measurements of this sort close to the monitors in conditions close to free field, it is normal to use a free-field, 0° incidence, measuring microphone pointing at the loudspeaker (this tells you most about the direct sound from the monitor itself). Further from the monitors the measurement will include more influence from the room and it is more correct to use a pressure microphone pointing upwards, or a small diaphragm measuring microphone whose response is not so direction dependent. The latter approach is likely to be closer to what people will hear in the room, although this depends on listening distance and loudspeaker directivity.

Measurement methods using time-delay spectrometry or maximum length sequence signal analysis (MLSSA) techniques enable one to look at the impulse response of the direct signal and the room separately, and one can 'window' the impulse to observe different time periods after the direct sound. A window that includes the direct arrival and a portion of the reflected sound will give an indication of the effects of early reflection interaction at the listening position. Such an impulse response can be transformed into the frequency domain using an FFT (fast Fourier transform) to view the result as a frequency spectrum.

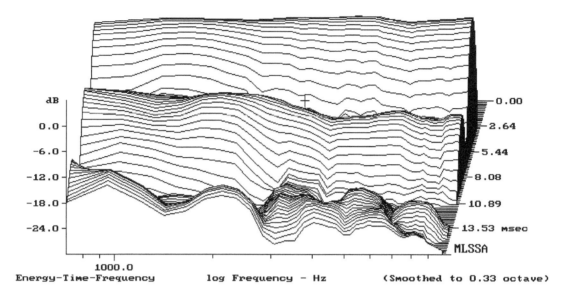

dB

0.0

-6.0

-12.0

-18.0

-24.0

0.00
2.64
5.44
8.08
10.89
13.53 msec

MLSSA

1000.0

Energy-Time-Frequency   log Frequency - Hz   (Smoothed to 0.33 octave)

Figure 5.9 Example of a waterfall plot from the MLSSA acoustic analyser showing time (front–back), frequency (left–right) and amplitude (vertical) of reflections after direct sound in a listening room.

Various time windows can be superimposed upon each other to create a so-called 'waterfall plot' that enables one to see the changing spectrum of the signal over time. This is useful for identifying the source of any particularly prominent reflections, and the effect on the resulting frequency spectrum (see Figure 5.9).

In recent years a number of so-called 'room correction' algorithms have been developed that attempt to measure the monitor chain, including the response of the room at the listening position, and calculate a suitable digital filter to correct the response (e.g. Genereux, 1992). Such systems normally include a proportion of the room reflection impulse response in the measured signal, so that the room interaction with the direct signal is calculated. The degree to which peaks and notches in the response should be completely ironed out is a matter for experimentation, as it is sometimes found that complete elimination of narrow, deep notches, for example leads to clipping of the monitor system or a dreadful response at another place in the room. Some form of spatial average is almost certainly needed, involving measuring the response over a range of positions and averaging them.

In the film industry, monitors are aligned to the so-called 'X' curve, which rolls off the HF content of the monitor output. An approximation to this is given in Figure 5.10 (it is modified according to room volume, and should be measured in the far field using pink noise and a small measuring microphone).

Figure 5.10 Typical shape of the film monitoring 'X' curve (depending on room volume).

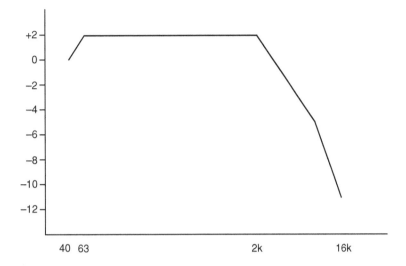

Because of the HF roll off in the monitor chain, HF boost often ends up being applied during recording, which can make film sound excessively toppy when replayed on flatter monitoring systems. As Holman (1999) explains, this can be removed in home systems using a process called re-equalisation, and this is employed in Home THX systems. Holman also provides some useful practical guidelines about the equalisation and level alignment of monitoring systems for 5.1 surround, including bass management. He recommends the use of spatial averaging for monitor equalisation, so that the effects of individual room modes and precise position dependency are reduced.

5.5 Virtual control room acoustics and monitors

Interested readers are referred to Section 3.2.9 for a discussion of systems that attempt to recreate a 'virtual' monitoring environment over headphones, using binaural technology. Such a solution could be particularly useful in monitoring environments such as OB trucks or small rooms where loudspeaker surround monitoring is not practical.

References

Alton-Everest, F. (1994). *The Master Handbook of Acoustics*. TAB Books.
Bech, S. (1995). Perception of reproduced sound: audibility of individual reflections in a complete sound field, II. Presented at *99th AES Convention, New York,* Preprint 4093.
Bech, S. (1998). Spatial aspects of reproduced sound in small rooms, *J. Acoust. Soc. Amer.,* **103**, pp. 434–445.

Bell, D. (2000). Surround sound studio design. *Studio Sound*, **42**, 7, July, pp. 55–58.

Genereux, R. (1992). Adaptive filters for loudspeakers and rooms. Presented at *93rd AES Convention, San Francisco*, Preprint 3375. Audio Engineering Society.

Griesinger, D. (1997). Spatial impression and envelopment in small rooms. Presented at *AES 103rd Convention, New York, September 26–29*. Preprint 4638. Audio Engineering Society.

Holman, T. and Zacharov, N. (2000). Comments on 'subjective appraisal of loudspeaker directivity for multichannel reproduction' (in Letters to the Editor). *J. Audio Eng. Soc.*, **48**, 4, pp. 314–321.

Holman, T. (1999). *5.1 Surround Sound: Up and Running*. Focal Press, Oxford and Boston.

Kügler, C. and Theile, G. (1992). Loudspeaker reproduction: study on the subwoofer concept. Presented at *92nd AES Convention, Vienna, 24–27 March*. Preprint 3335. Audio Engineering Society.

Munro, A. (1999). Cinema from the inside. *Studio Sound*, July, pp. 88–89.

Newell, P. (1995). *Studio Monitoring Design*. Focal Press, Oxford and Boston.

Newell, P. (1998). A matter of quality. *Studio Sound*, December, pp. 86–87.

Newell, P. (2000). Fold-down or melt-down? The case of the phantom speaker. *Audio Media*, January, pp. 93–100.

Nousaine, T. (1997). Multiple subwoofers for home theatre. Presented at *103rd AES Convention, New York, 26–29 September*. Preprint 4553. Audio Engineering Society.

Toole, F. and Olive, S. (1989). The detection of reflections in typical rooms. *J. Audio Eng. Soc.*, **37**, 7/8, pp. 539–553.

Varla, A. *et al.* (1999). Design of rooms for multichannel audio monitoring. In *Proceedings of the AES 16th International Conference: Spatial Sound Reproduction, Rovaniemi, 10–12 April*, pp. 523–531. Audio Engineering Society.

Voelker, E. (1998). Acoustics in control rooms – that recurring, burdensome subject. Presented at *105th AES Convention, San Francisco, 26–29 September*. Preprint 4832.

Zacharov, N., Bech, S. and Meares, D. (1998). The use of subwoofers in the context of surround sound reproduction. *J. Audio Eng. Soc.*, **46**, 4, pp. 276–287.

Zacharov, N. and Bech, S. (2000). Multichannel level alignment, part IV: the correlation between physical measures and subjective level calibration. Presented at *109th AES Convention, Los Angeles, 25–29 September*, Preprint 5241. Audio Engineering Society.

6 Two- and three-channel recording techniques

This chapter is concerned with basic recording techniques for two- and three-channel spatial audio systems – in other words the principles of 'conventional stereo' with a small extension to three front channels (in fact no more than a step back in history in some ways). Principles of two-channel recording technique are covered because they are a useful starting point for the understanding of the surround sound techniques covered in the chapter following this.

6.1 Science versus aesthetics in spatial recording

Some issues relating to the aims of spatial recording were introduced in Chapter 1, and they surface again here. The primary objects of discussion are the aesthetic standpoint of the mixing engineer, the nature of the programme material being recorded and the technical means by which those aesthetic aims are fulfilled. Recording 'engineering' is both an art and a science, and is not only about accurate acoustic rendering of natural sources and spaces.

6.1.1 Soundfield capture and reconstruction

Although possibly a somewhat artificial distinction, some recording methods are often classified as 'purist' techniques, usually to distinguish them as involving minimalist microphone arrays that capture the acoustic characteristics at the microphone position with reasonable spatial and timbral accuracy.

Conventional two-channel stereo represents a major improvement over mono in terms of spatial reproduction fidelity. It is possible to create phantom images between the loudspeakers and to create a sense of spaciousness in reproduction. Even so, no one in their right minds would claim that this is accurate sound field reconstruction since the loudspeakers are intended to be located in front of the listener and sources can only be imaged reliably within the angle subtended by the loudspeakers. All the reverberation in the reproduced sound (which originally came from all around) is reproduced from loudspeakers in front of the listener, in other words it comes from a direction similar to that of the sources. Although this can create a convincing illusion of the envelopment and spaciousness of the original environment, that is just what it is – a believable psychoacoustic illusion that depends on the technique of the recording engineer and the quality of the reproducing system.

In natural listening we generally expect sound to come from all around us, although in listening to music we generally orientate ourselves so that the main sources are in front of us and the reverberation from the environment comes from all around. Despite the fact that recent surround and 3D sound systems enable a distinctly enhanced spatial fidelity to be achieved in reproduction, there is still some way to go before accurate sound field reconstruction can be achieved in all three dimensions for listeners sitting in multiple positions in relation to the loudspeakers. Even if it were possible, most research suggests that this would need considerably larger numbers of loudspeakers than we currently have in standard systems (many thousands were suggested by Michael Gerzon), or that listeners would have to be restricted in their positions. Work in the Netherlands at Delft University, dealing with wavefield synthesis using large numbers of channels, gets closest to this (e.g. Berkhout *et al.* 1992). Binaural approaches, discussed in Chapter 3, also make accurate three-dimensional rendering a possibility but limit the range of listening positions the listener can occupy or require them to wear headphones.

6.1.2 Creative illusion and 'believability'

It would be reasonable to surmise that in most practical circumstances, for mainstream consumer applications, we will be dealing with the business of creating believable illusions for some time to come. In other words, we will need to create the *impression* of natural spaces, source positions, depth, size and so on, without necessarily being able to replicate the exact sound pressure and velocity vectors that would be needed at each listening position to recreate a sound field accurately. One must remember that listeners rarely sit in the optimum listening position, and often like to move around while listening. While it may be possible to achieve greater spatial accuracy using headphone reproduction, headphones are not always a practical or desirable form of monitoring.

Günther Theile puts some of the issues relating to 'natural' sound recording quite well in a recent paper on multichannel microphone techniques (Theile, 2000):

> What does optimum naturalness mean? The simplest answer would be: the reproduced sound image must be as identical as possible with the original sound image. This definition appears to be problematic because identity can definitely not be required, in principle, as a goal for optimising the stereophonic technique. Identity may conceivably be appropriate for dummy-head stereophony or wavefield synthesis, or perhaps for the reproduction of a speaker's voice through loudspeakers, but it is appropriate to a limited extent only for the reproduction of the sound of a large orchestra through loudspeakers. Artistic intentions of the sound engineer, aesthetic irregularities in the orchestra, poor recording conditions in the concert hall, as well as the necessity of creating a sound mix 'suitable for a living room' with respect to practical constraints (poor listening conditions, reduced dynamic, downward compatibility) – in other words, artistic sound design requirements and the essential problems of loudspeaker stereophony actually result in a deviation from identity.

The desired natural stereophonic image should therefore meet two requirements: it should satisfy aesthetically and it should match the tonal and spatial properties of the original sound at the same time. Both requirements will undoubtedly be contradictory in many situations. However, the compromise, namely, optimisation by the sound engineer, will be the better, the more flexible the stereophonic recording technique is.

Theile summarises the attributes of some different stereo formats in terms of their capabilities in these respects, as shown in Table 6.1.

In movie and television sound, an image is present that is normally projected in front of the listener. The primary attention

Table 6.1 Spatial capabilities of different stereo recording/reproduction systems (after Günther Theile)

	2/0 stereo	3/2 stereo	Dummy head
Horizontal direction	±30°	±30°, surround effects	Surround (unstable front)
Elevation	Not possible	Constraints?	Possible
Depth	Simulated	Constraints?	Possible
Near-head distance	Not possible	No?	Possible
Spatial impression	Simulated	Possible	Possible
Enveloping sources	Not possible	Constraints?	Possible

is therefore in front and the majority of sound is related to activity on the picture. Most dialogue sound is restricted to the centre. Effects in the rear channels are used sparingly to create an immersive and spatially interesting environment that complements the picture and does not contradict what the visual sense is telling the viewer/listener. Again we are dealing with a predominantly front-biased spatial mixing paradigm in which sounds to the rear and sides are used sparingly and mainly to imply an appropriate environmental context.

6.1.3 Applications and appropriate treatment

The spatial mixing paradigm employed in sound recording will be dictated by the needs of the programme material, the commercial imperatives of the project and the aesthetic preferences of the producer and sound mixer. The decisions about choice of spatial mixing paradigm will be dictated by 'appropriateness'. Often, as mentioned in Chapter 1, no natural acoustic environment is intended to be implied in the recording, it being a purely artificial creation – a form of 'acoustic fiction'. In such cases, it might be argued, anything goes and there are no rules about how sound images should be created and how the sound space should be used. While this argument has some weight, total aesthetic freedom can also lead to poor results, and many creative people have found historically that some limits or restrictions on what they can do leads to a more disciplined and ultimately more structured product.

Classical music recording is likely to adhere more to the soundfield reconstruction school of thought than pop recording, for very good reasons, although the degree to which the natural acoustic experience can be emulated in practice is questionable. In classical and other forms of 'live acoustic' recording such as some jazz, an acoustic environment is part of the sound to be captured. Indeed the venue is usually chosen for its favourable acoustic characteristics. Consequently the spatial recording technique will be designed as far as possible to capture the spatial characteristics not only of the musical sources but of the venue. As will be seen below, though, there are serious limitations to the capacity of even four or five channel spatial audio formats to convey accurate acoustic cues for three-dimensional rendering, although a distinct improvement over two-channel stereo is possible.

Many other forms of recording, though, such as pop music, television and film, rely on more artificial forms of balancing that depend on panning techniques and artificial reverberation for their spatial effect, sometimes with natural acoustic content mixed in. The primary debate here, in relation to surround sound formats, is how to utilise the spatial capabilities of the surround channels in a conventional five-channel array, given the spacing and angle between the loudspeakers. The limitations this format places on the creative options for spatial mixing are the boundaries limiting complete artistic flexibility mentioned above. If discrete loudspeaker-feed panning and recording techniques are to be used then certain uses of these channels are simply not sensible, as will be discussed below. Those that promote the use of Ambisonics (discussed in Chapter 4) would probably say that this is the natural consequence of settling on a crude 'cinema-style' surround format for all purposes, and that an approach based more strongly on psychoacoustic principles and soundfield representation would free the user from such creative limitations.

6.2 Two-channel microphone techniques

This section contains a review of basic two channel microphone techniques, upon which many spatial recording techniques are based. Panned spot microphones are often mixed in to the basic stereo image created by such techniques. For a detailed coverage of microphone principles and types the reader is referred to *The Microphone Engineering Handbook*, edited by Michael Gayford (Gayford, 1994).

6.2.1 Coincident pair principles

The coincident-pair incorporates two directional capsules that may be angled over a range of settings to allow for different configurations and operational requirements. The pair can be operated in either the LR (sometimes known as 'XY') or MS

Figure 6.1 A typical coincident stereo microphone (Neumann SM69).

modes (see Section 3.1.4), and a matrixing unit is sometimes supplied with microphones which are intended to operate in the MS mode in order to convert the signal to LR format for recording. The directional patterns (polar diagrams) of the two microphones need not necessarily be figure-eight, although if the microphone is used in the MS mode the S capsule must be figure-eight (see below). Directional information is encoded solely in the level differences between the capsule outputs, since the two capsules are mounted physically as close as possible, and therefore there are no phase differences between the outputs except at the highest frequencies where inter-capsule spacing may become appreciable in relation to the wavelength of sound.

Coincident pairs can be manufactured as integrated stereo microphones, such as the one shown in Figure 6.1. They are normally mounted vertically in relation to the sound source, so that the two capsules are angled to point left and right (see Figure 6.2). The choice of angle depends on the polar response of the capsules used, since most stereo mikes allow for either capsule to be switched through a number of pickup patterns between figure-eight and omnidirectional. A coincident pair of figure-eight microphones at 90° provides good correspondence between the actual angle of the source and the apparent position of the virtual image when reproduced on loudspeakers, but there are also operational disadvantages to the figure-eight pattern in some cases, such as the amount of reverberation pickup.

Figure 6.3 shows the polar pattern of a coincident pair using figure-eight mikes. Firstly, it may be seen that the fully-left position corresponds to the null point of the right capsule's

Figure 6.2 A coincident pair's capsules are oriented so as to point left and right of the centre of the sound stage.

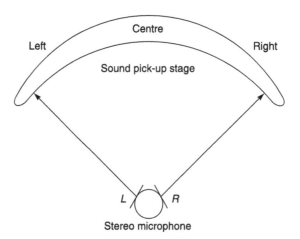

Figure 6.3 Polar pattern of a coincident pair using figure-eight microphones.

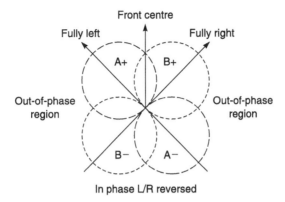

pickup. This is the point at which there will be maximum level difference between the two capsules. The fully left position also corresponds to the maximum pickup of the left capsule but it does not always do so in other stereo pairs. As a sound moves across the sound stage from left to right it will result in a gradually decreasing output from the left mike, and an increasing output from the right mike. Since the microphones have cosine responses, the output at 45° off axis is √2 times the maximum output, or 3 dB down in level, thus the takeover between left and right microphones is smooth for music signals.

The second point to consider with this pair is that the rear quadrant of pickup suffers a left–right reversal, since the rear lobes of each capsule point in the opposite direction to the front. This is important when considering the use of such a microphone in situations where confusion may arise between sounds picked up on the rear and in front of the mike, such as in television sound where the viewer can also see the positions of sources. The third point is that pickup in both side quadrants results in out-of-phase signals between the channels, since a source further round than 'fully left' results in pickup by both the negative lobe of the right capsule and the positive lobe of the left capsule. There is thus a large region around a crossed pair of figure-eights that results in out-of-phase information, this information often being reflected or reverberant sound. Any sound picked up in this region will suffer cancellation if the channels are summed to mono, with maximum cancellation occurring at 90° and 270°, assuming 0° as the centre-front.

The operational advantages of the figure-eight pair are the crisp and accurate phantom imaging of sources, together with a natural blend of ambient sound from the rear. Some cancellation of ambience may occur, especially in mono, if there is a lot of

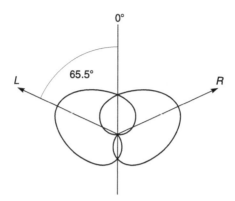

Figure 6.4 A coincident pair of cardioid microphones should theoretically be angled at 131°, but deviations either side of this may be acceptable in practice.

reverberant sound picked up by the side quadrants. Disadvantages lie in the large out-of-phase region, and in the size of the rear pickup which is not desirable in all cases and is left–right reversed. Stereo pairs made up of capsules having less rear pickup may be preferred in cases where a 'drier' or less reverberant balance is required, and where frontal sources are to be favoured over rear sources. In such cases the capsule responses may be changed to be nearer the cardioid pattern, and this requires an increased angle between the capsules to maintain good correlation between actual and perceived angle of sources.

The cardioid crossed pair shown in Figure 6.4 is angled at approximately 131°, although angles of between 90° and 180° may be used to good effect depending on the width of the sound stage to be covered. At an angle of 131° a centre source is 65.5° off-axis from each capsule, resulting in a 3 dB drop in level compared with the maximum on-axis output (the cardioid mike response is equivalent to $0.5\,(1 + \cos\vartheta)$, where ϑ is the angle off-axis of the source, and thus the output at 65.5° is $\sqrt{2}$ times that at 0°). A departure from the theoretically correct angle is often necessary in practical situations, and it must be remembered that the listener will not necessarily be aware of the 'correct' location of each source, neither may it matter that the true and perceived positions are different. A pair of 'back-to-back' cardioids has often been used to good effect (see Figure 6.5), since it has a simple MS equivalent of an omni and a figure-eight, and has no out-of-phase region. Although the maximum level difference between the channels is at 90° off-centre there will in fact be a satisfactory level difference for a phantom image to appear fully left or right at a substantially smaller angle than this.

With any coincident pair, fully left or fully right corresponds to the null point of pickup of the opposite channel's microphone,

Figure 6.5 Back-to-back cardioids have been found to work well in practice and should have no out-of-phase region.

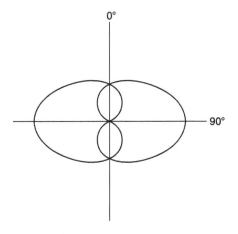

although psychoacoustically this point may be reached before the maximum level difference is arrived at. This also corresponds to the point where the M signal equals the S signal (where the sum of the channels is the same as the difference between them). As the angle between the capsules is made larger, the angle between the null points will become smaller (see Figure 6.6). Operationally, if one wishes to widen the reproduced sound stage one will widen the angle between the microphones which

Figure 6.6 There is a difference between the acceptance angle of a stereo pair (the angle between the pickup null points) and the angle between the capsules. As the angle between the capsules is increased (as shown in (b)) the acceptance angle decreases, thus widening the stereo image.

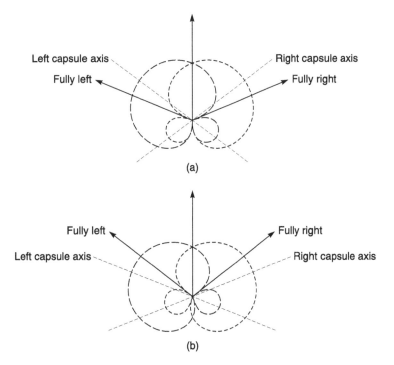

is intuitively the right thing to do. This results in a narrowing of the angle between fully left and fully right, so sources which had been, say, half left in the original image will now be further towards the left. A narrow angle between fully left and fully right results in a very wide sound stage, since sources have only to move a small distance to result in large changes in reproduced position. This corresponds to a wide angle between the capsules.

Further coincident pairs are possible using any polar pattern between figure-eight and omni, although the closer that one gets to omni the greater the required angle to achieve adequate separation between the channels. The hypercardioid pattern is often chosen for its smaller rear lobes than the figure-eight, allowing a more distant placement from the source for a given direct-to-reverberant ratio (although in practice hypercardioid pairs tend to be used closer to make the image width similar to that of a figure-eight pair). Since the hypercardioid pattern lies between figure-eight and cardioid, the angle required between the capsules is correctly around 110°.

Psychoacoustic requirements introduced earlier in the book suggest the need for an electrical narrowing of the image at high frequencies in order to preserve the correct angular relationships between low and high frequency signals, although this is rarely implemented in practice with coincident pair recording. A further consideration to do with the theoretical versus the practical is that although microphones tend to be referred to as having a particular polar pattern, this pattern is unlikely to be consistent across the frequency range and this will have an effect on the stereo image. Cardioid crossed pairs should theoretically exhibit no out-of-phase region (there should be no negative rear lobes), but in practice most cardioid capsules become more omni at LF and narrower at HF, so some out-of-phase components may be noticed in the HF range while the width may appear too narrow at LF. Attempts have been made to compensate for this in some stereo microphone designs.

XY or LR coincident pairs in general have the possible disadvantage that central sounds are off-axis to both mikes, perhaps considerably so in the case of crossed cardioids. This may result in a central signal with a poor frequency response and possibly an unstable image if the polar response is erratic. Whether or not this is important depends on the importance of the central image in relation to that of offset images, and will be most important in cases where the main source is central (such as in television, with dialogue). In such cases the MS technique described in the next section is likely to be more appropriate, since central sources will be on-axis to the M microphone. For music recording it would be hard to say whether central sounds are any more important than offset sources, so either technique may be acceptable.

6.2.2 Using MS processing on coincident pairs

Although some stereo microphones are built specifically to operate in the MS mode, it is possible to take any coincident pair capable of at least one capsule being switched to figure-eight, and orientate it so that it will produce suitable signals. The S component (being the difference between left and right signals) is always a sideways-facing figure-eight with its positive lobe facing left. The M (middle) component may be any polar pattern facing to the centre-front, although the choice of M pattern depends on the desired equivalent pair, and will be the signal that a mono listener would hear. True MS mikes usually come equipped with a control box that matrixes the MS signals to LR format if required. A control for varying S gain is often provided as a means of varying the effective acceptance angle between the equivalent LR pair.

MS signals are not suitable for direct stereo monitoring, they are sum and difference components and must be converted to a conventional loudspeaker format at a convenient point in the production chain. The advantages of keeping a signal in the MS format until it needs to be converted will be discussed below, but the major advantage of pickup in the MS format is that central signals will be on-axis to the M capsule, resulting in the best frequency response. Furthermore, it is possible to operate an MS mike in a similar way to a mono mike, which may be useful in television operations where the MS mike is replacing a mono mike on a pole or in a boom. Hibbing (1989), amongst others, points to the reduced audible effects of variations in microphone polar pattern with frequency when using the MS pickup technique.

To see how MS and LR pairs relate to each other, and to draw some useful conclusions about stereo width control, it is informative to consider a coincident pair of figure-eight mikes again. For each MS pair there is an LR equivalent. The polar pattern of the LR equivalent to any MS pair may be derived by plotting the level of $(M + S)/2$ and $(M - S)/2$ for every angle around the pair. Taking the MS pair of figure-eight mikes shown in Figure 6.7, it may be seen that the LR equivalent is simply another pair of figure-eights, but rotated through 45°. Thus the correct MS arrangement to give an equivalent LR signal where both

Figure 6.7 Every coincident pair has an MS equivalent. The conventional left–right arrangement is shown in (a), and the MS equivalent in (b).

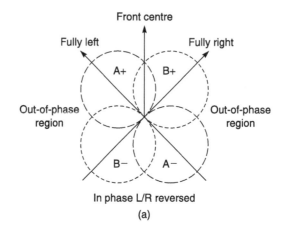

(a)

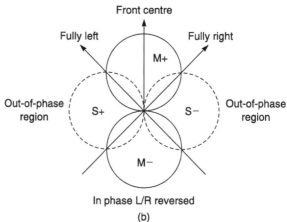

(b)

'capsules' are orientated at 45° to the centre-front (the normal arrangement) is for the M capsule to face forwards and the S capsule to face sideways.

A number of interesting points arise from a study of the LR/MS equivalence of these two pairs, and these points apply to all equivalent pairs. Firstly, fully left or right in the resulting stereo image occurs at the point where S = M (in this case at 45° off centre). This is easy to explain, since the fully left point is the point at which the output from the right capsule is zero. Therefore M = L + 0, and S = L − 0, both of which equal L. Secondly, at angles of incidence greater than 45° off centre in either direction the two channels become out-of-phase, as was seen above, and this corresponds to the region in which S is greater than M. Thirdly, in the rear quadrant where the signals are in phase again, but left–right reversed, the M signal is greater than S again. The relationship between S and M levels, therefore, is an excellent guide to the phase relationship between the equivalent LR signals. If S is lower than M, then the LR signals will be in phase. If S = M, then the source is either fully left or right, and if S is greater than M, then the LR signals will be out-of-phase.

To show that this applies in all cases, and not just that of the figure-eight pair, look at the MS pair in Figure 6.8 together with its LR equivalent. This MS pair is made up of a forward-facing cardioid and a sideways-facing figure-eight (a popular arrangement). Its equivalent is a crossed pair of hypercardioids, and again the extremes of the image (corresponding to the null points of the LR hypercardioids) are the points at which S equals M. Similarly, the signals go out-of-phase in the region where S is greater than M, and come back in phase again for a tiny angle

Figure 6.8 The MS equivalent of a forward facing cardioid and sideways figure-eight, as shown in (a), is a pair of hypercardioids whose effective angle depends on S gain, as shown in (b).

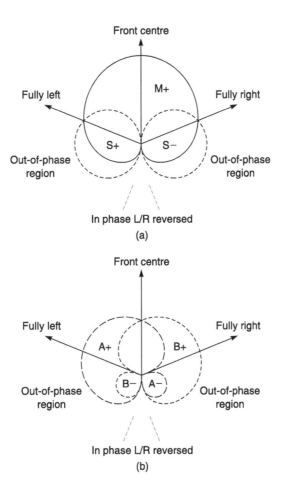

round the back, due to the rear lobes of the resulting hypercardioids. Thus the angle of acceptance (between fully left and fully right) is really the frontal angle between the two points on the MS diagram where M equals S.

Now, consider what would happen if the gain of the S signal was raised (imagine expanding the lobes of the S figure-eight in Figure 6.8). The result of this would be that the points where S equalled M would move inwards, making the acceptance angle smaller. As explained earlier, this results in a wider stereo image, since off-centre sounds will become closer to the extremes of the image, and is equivalent to increasing the angle between the equivalent LR capsules. Conversely, if the S gain is reduced, the points at which S equals M will move further out from the centre, resulting in a narrower stereo image, equivalent to decreasing the angle between the equivalent AB capsules. This

Figure 6.9 Polar patterns of the Neumann RSM191i microphone. (a) M capsule, (b) S capsule, (c) LR equivalent with –6 dB S gain, (d) 0 dB S gain, (e) +6 dB S gain.

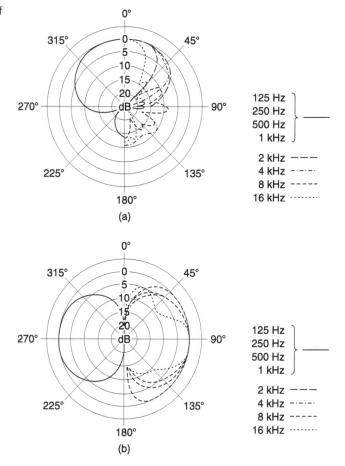

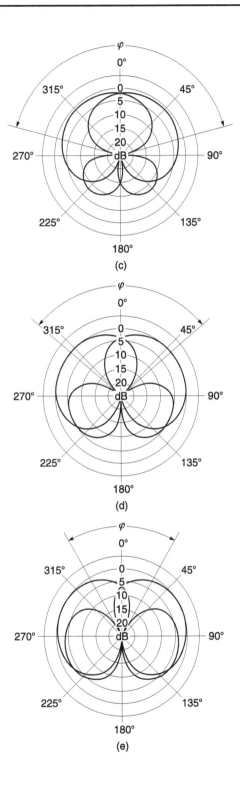

(c)

(d)

(e)

helps to explain why Blumlein-style shufflers work by processing the MS equivalent signals of stereo pairs, as one can change the effective stereo width of pairs of signals, and this can be made frequency dependent if required.

This is neatly exemplified in a commercial example, the Neumann RSM 191i, which is an MS mike in which the M capsule is a forward-facing short shotgun mike with a polar pattern rather like a hypercardioid. The polar pattern of the M and S capsules and the equivalent LR pair is shown in Figure 6.9 for three possible gains of the S signal with relation to M (–6 dB, 0 dB and +6 dB). It will be seen that the acceptance angle (ϑ) changes from being large (narrow image) at –6 dB, to small (wide image) at +6 dB. Changing the S gain also affects the size of the rear lobes of the LR equivalent. The higher the S gain the larger the rear lobes. Not only does S gain change stereo width, it also affects rear pickup, and thus the ratio of direct to reverberant sound.

Any stereo pair may be operated in the MS configuration, simply by orientating the capsules in the appropriate directions and switching them to an appropriate polar pattern, but certain microphones are dedicated to MS operation simply by the physical layout of the capsules. There is not space here to show all the possible MS pairs and their equivalents, but a comprehensive review may be found in Dooley and Streicher (1982).

6.2.3 Operational considerations with coincident pairs

The control of S gain is an important tool in determining the degree of width of a stereo sound stage, and for this reason the MS output from a microphone might be brought (unmatrixed) into a mixing console, so that the engineer has control over the width. This in itself can be a good reason for keeping a signal in MS form during the recording process, although M and S can easily be derived at any stage using a conversion matrix.

Although some mixers have MS matrixing facilities on board, the diagram in Figure 6.10 shows how it is possible to derive an LR mix with variable width from an MS microphone using three channels on a mixer without using an external MS matrix. M and S outputs from the microphone are fed in phase through two mixer channels and faders, and a post-fader feed of S is taken to a third channel line input, being phase-reversed on this channel. The M signal is routed to both left and right mix buses (panned centrally), whilst the S signal is routed to the left mix bus (M + S = 2L) and the –S signal (the phase-reversed version) is routed

Figure 6.10 An LR mix with variable width can be derived from an MS microphone connected to three channels of a mixer as shown. The S faders should be ganged together and used as a width control.

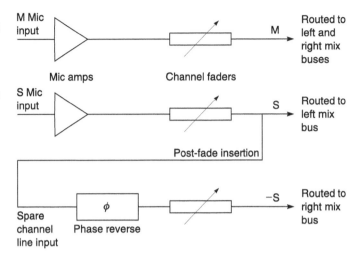

to the right mix bus (M − S = 2R). It is important that the gain of the −S channel is matched very closely with that of the S channel. (A means of deriving M and S from an LR format input is to mix L and phase-reversed R together to get S, and without the phase reverse to get M.)

Outdoors, coincident pairs will be susceptible to wind noise and rumble, as they incorporate velocity-sensitive capsules, which always give more problems in this respect than omnis. Most of the interference will reside in the S channel, since this has always a figure-eight pattern, and thus would not be a problem to the mono listener. Similarly, physical handling of the stereo microphone, or vibration picked up through a stand, will be much more noticeable than with pressure microphones. Coincident pairs should not generally be used close to people speaking, as small movements of their heads can cause large changes in the angle of incidence, leading to considerable movement in their apparent position in the sound stage.

6.2.4 Near-coincident microphone configurations

It is generally admitted that the vector summation theory of sounds from loudspeakers having only level differences between them only holds true for continuous low frequency sounds up to around 700 Hz, where head-shadowing effects begin to take over (see Chapter 3). Coincident pairs work by creating level differences only between channels. The process by which transient sounds and HF sounds are handled by such arrays are not so obviously analysable, although various attempts have been made to show how such sources could be localised with coincident pair signals on loudspeakers.

'Near coincident' pairs of directional microphones introduce small additional timing differences which may help in the localisation of transient sounds and increase the spaciousness of a recording, while at the same time remaining nominally coincident at low frequencies and giving rise to suitable amplitude differences between the channels. Headphone compatibility is also quite good owing to the microphone spacing being similar to ear spacing. The family of near-coincident (or closely spaced) techniques relies on a combination of time and level differences between the channels that can be traded off for certain widths of sound stage and microphone pattern.

Subjective evaluations often seem to show good results for such techniques. One comprehensive subjective assessment of stereo microphone arrangements, performed at the University of Iowa (Cross, 1985), consistently resulted in the near-coincident pairs scoring amongst the top few performers for their sense of 'space' and realism. Lipshitz (1986) attributed these effects to 'phasiness' at high frequencies (which some people may like, nonetheless), and argued that truly coincident pairs were preferable. Gerzon (1986) suggested that a very small spacing of crossed cardioid mikes (about 5 cm) could actually compensate for the phase differences introduced when 'spatial equalisation' was used (the technique described earlier of increasing LF width relative to HF width by introducing equalisation into the S channel).

A number of examples of near-coincident pairs exist as 'named' arrangements, although, as Williams (1987) has shown, there is a whole family of possible near-coincident arrangements using combinations of spacing and angle. Some near-coincident pairs of different types, based on the 'Williams curves' are given in Table 6.2. The so-called 'ORTF pair' is an arrangement of two cardioid mikes, deriving its name from the organisation which first adopted it (the Office de Radiodiffusion-Television Française). The two mikes are spaced apart by 170 mm, and angled at 110°. The 'NOS' pair (Nederlande Omroep Stichting, the Dutch Broadcasting Company), uses cardioid mikes spaced apart by 300 mm and angled at 90°. Figure 6.11 illustrates these two pairs, along with a third pair of figure-eight microphones spaced apart by 200 mm, which has been called a 'Faulkner' pair, after the British recording engineer who first adopted it. This latter pair has been found to offer good image focus on a small-to-moderate-sized central ensemble with the mikes placed further back than would normally be expected.

Figure 6.11 Near-coincident pairs: (a) ORTF, (b) NOS, (c) Faulkner.

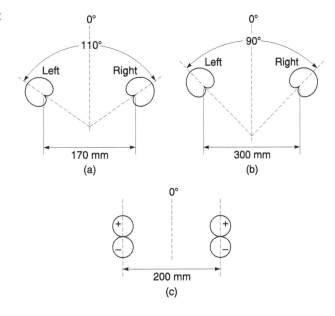

Table 6.2 Some near coincident pairs based on the 'Williams curves'

Designation	Polar pattern	Mike angle	Spacing	Recording angle
NOS	Cardioid	±45°	30 cm	80°
RAI	Cardioid	±50°	21 cm	90°
ORTF	Cardioid	±55°	17 cm	95°
DIN	Cardioid	±45°	20 cm	100°
–	Omni	0°	50 cm	130°
–	Omni	0°	35 cm	160°

The success of near-coincident arrays in practice may be attributed to their compromise nature. Two-channel stereo almost demands a compromise to be struck between imaging accuracy and spaciousness in the microphone technique, since neither alone is capable of providing the believable illusion of natural spatial acoustics that most people want to achieve. Two-channel techniques that provide excellent phantom imaging accuracy in the angle between the loudspeakers do not always give rise to the interchannel differences required for a strong sense of envelopment and spaciousness. More widely spaced microphone techniques that achieve the latter do not necessarily achieve the former so well.

6.2.5 Pseudo-binaural techniques

Binaural techniques could be classed as another form of near coincident technique. The spacing between the omni micro-

Figure 6.12 The Schoeps KFM6U microphone consists of two pressure microphones mounted on the surface of a sphere. (Courtesy of Schalltechnik Dr.-Ing. Schoeps GmbH).

phones in a dummy head is not great enough to fit any of the Williams models described above for near-coincident pairs, but the shadowing effect of the head makes the arrangement more directional at high frequencies. Low frequency width is likely to need increasing to make the approach more loudspeaker-compatible, as described in Section 3.2.6.

As mentioned in Sections 3.1.1 and 3.2.6, 'head-related' or binaural signals may be used directly as loudspeaker signals if one agrees with Theile's association model of spatial reproduction. The Schoeps KFM6U microphone, pictured in Figure 6.12, was designed as a head-sized sphere with pressure microphones mounted on the surface of the sphere, equalised for a flat response to frontal incidence sound and suitable for generating signals that could be reproduced on loudspeakers. This is in effect a sort of dummy head without ears. Dummy heads also exist that have been equalised for a reasonably natural timbral quality on loudspeakers, such as the Neumann KU100. The use of unprocessed dummy head techniques for stereo recording intended for loudspeakers has found favour with some recording engineers because they claim to like the spatial impression created, although others find the stereo image somewhat unfocused or vague.

6.2.6 Spaced microphone configurations

Spaced arrays have a historical precedent for their usage, since they were the first to be documented (in the work of Clement Ader at the Paris Exhibition in 1881), were the basis of the Bell Labs' stereo systems in the 1930s, and have been widely used since then. They are possibly less 'correct' theoretically, from a standpoint of sound field representation, but they can provide a number of useful spatial cues that give rise to believable illusions of natural spaces. Many recording engineers prefer spaced arrays because the omni microphones often used in such arrays tend to have a flatter and more extended frequency response than their directional counterparts, although it should be noted that spaced arrays do not have to be made up of omni mikes (see below).

Spaced arrays rely principally on the precedence effect, discussed in Chapter 2. The delays that result between the channels tend to be of the order of a number of milliseconds. With spaced arrays the level and time difference resulting from a source at a particular left–right position on the sound stage will depend on how far the source is from the microphones (see Figure 6.13), with a more distant source resulting in a much smaller delay and level difference. In order to calculate the time

Figure 6.13 With spaced omnis a source at position X results in path lengths d_1 and d_2 to each microphone respectively, whilst for a source in the same *LR* position but at a greater distance (source Y) the path length difference is smaller, resulting in a smaller time difference than for X.

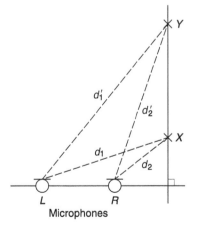

and level differences which will result from a particular spacing it is possible to use the following two formulae:

$$\Delta t = (d_1 - d_2)/c; \qquad \Delta L = 20 \log_{10}(d_1/d_2)$$

where Δt is the time difference and ΔL the pressure level difference which results from a source whose distance is d_1 and d_2 respectively from the two microphones, and c is the speed of sound (340 m/s).

When a source is very close to a spaced pair there may be a considerable level difference between the microphones, but this will become small once the source is more than a few metres distant. The positioning of spaced microphones in relation to a source is thus a matter of achieving a compromise between closeness (to achieve satisfactory level and time differences between channels), and distance (to achieve adequate reverberant information relative to direct sound). When the source is large and deep, such as a large orchestra, it will be difficult to place the microphones so as to suit all sources, and it may be found necessary to raise the microphones somewhat so as to reduce the differences in path length between sources at the front and rear of the orchestra.

Spaced microphone arrays do not stand up well to theoretical analysis when considering the imaging of continuous sounds, the precedence effect being related principally to impulsive or transient sounds. Because of the phase differences between signals at the two loudspeakers created by the microphone spacing, interference effects at the ears at low frequencies may in fact result in a contradiction between level and time cues at the ears. It is possible in fact that the ear on the side of the earlier signal may not experience the higher level, thus producing a confusing difference between the cues provided by impulsive sounds and those provided by continuous sounds. The lack of phase coherence in spaced-array stereo is further exemplified by phase inverting one of the channels on reproduction, an action which does not always appear to affect the image particularly, as it would with coincident stereo, showing just how uncorrelated the signals are. (This is most noticeable with widely spaced microphones.)

Accuracy of phantom image positioning is therefore lower with spaced arrays, although many convincing recordings have resulted from their use. Lipshitz argued in 1986 that the impression of spaciousness that results from the use of spaced arrays is in fact simply the result of phasiness and comb-filtering effects. Others suggest that there is a place for the spaciousness

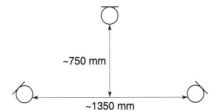

Figure 6.14 The classic 'Decca Tree' involves three omnis, with the centre microphone spaced slightly forward of the outer mikes.

~750 mm

~1350 mm

that results from spaced techniques, since the highly decorrelated signals which result from spaced techniques are also a feature of concert hall acoustics. Griesinger has often claimed informally that spacing the mikes apart by at least the reverberation radius (critical distance) of a recording space gives rise to adequate decorrelation between the microphones to obtain good spaciousness, and that this might be a suitable technique for ambient sound in surround recording. Mono compatibility of spaced pairs is variable, although not always as poor in practice as might be expected.

The so-called 'Decca Tree' is a popular arrangement of three spaced omnidirectional mikes. The name derives from the traditional usage of this technique by the Decca Record Company, although even that company did not adhere rigidly to this arrangement. A similar arrangement is described by Grignon (1949). Three omnis are configured according to the diagram in Figure 6.14, with the centre microphone spaced so as to be slightly forward of the two outer mikes, although it is possible to vary the spacing to some extent depending on the size of the source stage to be covered. The reason for the centre microphone and its spacing is to stabilise the central image which tends otherwise to be rather imprecise, although the existence of the centre mike will also complicate the phase relationships between the channels, thus exacerbating the comb-filtering effects that may arise with spaced pairs. The advance in time experienced by the forward mike will tend to solidify the central image, due to the precedence effect, avoiding the hole-in-the-middle often resulting from spaced pairs. The outer mikes are angled outwards slightly, so that the axes of best HF response favour sources towards the edges of the stage whilst central sounds are on-axis to the central mike.

A pair of omni outriggers are often used in addition to the tree, towards the edges of wide sources such as orchestras and choirs, in order to support the extremes of the sound stage that are at some distance from the tree (see Figure 6.15). This is hard to justify on the basis of any conventional imaging theory, and is

Figure 6.15 Omni outriggers may be used in addition to a coincident pair or Decca Tree, for wide sources.

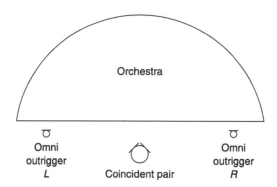

beginning to move towards the realms of multi-microphone pickup, but can be used to produce a commercially acceptable sound. Once more than around three microphones are used to cover a sound stage one has to consider a combination of theories, possibly suggesting conflicting information between the outputs of the different microphones. In such cases the sound balance will be optimised on a mixing console, subject to the creative control of the recording engineer.

Spaced microphones with either omnidirectional or cardioid patterns may be used in configurations other than the Decca Tree described above, although the 'tree' has certainly proved to be the more successful arrangement in practice. The precedence effect begins to break down for delays greater than around 40 ms, because the brain begins to perceive the two arrivals of sound as being discrete rather than integrated. It is therefore reasonable to assume that spacings between microphones which give rise to greater delays than this between channels should be avoided. This maximum delay, though, corresponds to a mike spacing of well over 10 metres, and such extremes have not proved to work well in practice due to the great distance of central sources from either microphone compared with the closeness of sources at the extremes, resulting in a considerable level drop for central sounds and thus a hole in the middle. Indeed if one looks back to Chapters 2 and 3, one can see that delays of only about 0.5–1.5 ms seem to be required to result in fully left or right phantom images, suggesting no need for wide microphone spacings on the basis of imaging requirements alone.

Dooley and Streicher (1982) have shown that good results may be achieved using spacings of between one-third and one-half of the width of the total sound stage to be covered (see Figure 6.16), although closer spacings have also been used to good effect. Bruel and Kjaer manufacture matched stereo pairs of omni

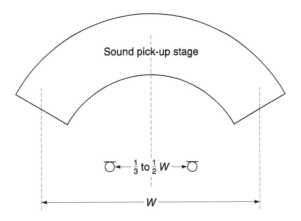

Figure 6.16 Dooley and Streicher's (1982) proposal for omni spacing.

Sound pick-up stage

$\frac{1}{3}$ to $\frac{1}{2}$ W

W

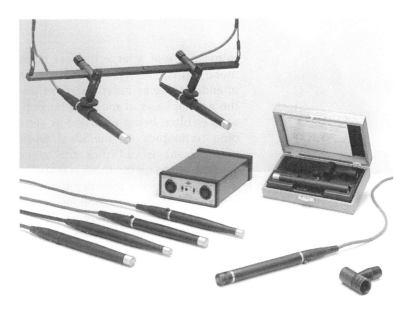

Figure 6.17 B&K omni microphones mounted on a stereo bar that allows variable spacing.

microphones together with a bar which allows variable spacing, as shown in Figure 6.17, and suggest that the spacing used is smaller than one-third of the stage width (they suggest between 5 cm and 60 cm, depending on stage width), their principal rule being that the distance between the microphones should be small compared with the distance from microphones to source.

6.3 Spot microphones and two-channel panning laws

We have so far considered the use of a small number of microphones to cover the complete sound stage, but it is also possible

Figure 6.18 Typical two-channel panpot law used in sound mixers.

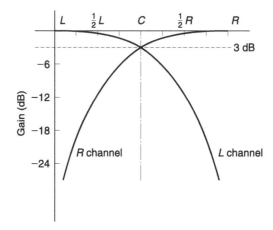

to make use of a large number of mono microphones or other mono sources, each covering a small area of the sound stage and intended to be as independent of the others as possible. This is the normal basis of most studio pop music recording, with the sources often being recorded at separate times using overdubbing techniques. In the ideal world, each mike in such an arrangement would pick up sound only from the desired sources, but in reality there is usually considerable spill from one to another. It is not the intention in this chapter to provide a full resumé of studio microphone technique, and thus discussion will be limited to an overview of the principles of multi-mike pickup as distinct from the more simple techniques described above.

In multi-mike recording each source feeds a separate channel of a mixing console, where levels are individually controlled and the mike signal is 'panned' to a virtual position somewhere between left and right in the sound stage. The pan control takes the monophonic signal and splits it two ways, controlling the proportion of the signal fed to each of the left and right mix buses. Typical pan control laws follow a curve which gives rise to a 3 dB drop in the level sent to each channel at the centre, resulting in no perceived change in level as a source is moved from left to right (see Figure 6.18). This has often been claimed to be due to the way signals from left and right loudspeakers sum acoustically at the listening position, which includes a diffuse field component of the room. The −3 dB panpot law is not correct if the stereo signal is combined electrically to mono, since the summation of two equal signal voltages would result in a 6 dB rise in level for signals panned centrally, and thus a

–6 dB law is more appropriate for mixers whose outputs will be summed to mono (e.g. radio and TV operations) as well as stereo, although this will then result in a drop in level in the centre for stereo signals. A compromise law of –4.5 dB is sometimes adopted by manufacturers for this reason.

Panned mono balances therefore rely on channel level differences, separately controlled for each source, to create phantom images on a synthesised sound stage, with relative level between sources used to adjust the prominence of a source in a mix. Time delay is hardly ever used as a panning technique, for reasons of poor mono compatibility and technical complexity. Artificial reverberation may be added to restore a sense of space to a multi-mike balance. Source distance can be simulated by the addition of reflections and reverberation, as well as by changes in source spectrum and overall level (e.g. HF roll off can simulate greater distance).

It is common in classical music recording to use close mikes in addition to a coincident pair or spaced pair in order to reinforce sources which appear to be weak in the main pickup, these close mikes being panned as closely as possible to match the true position of the source. The results of this are variable and can have the effect of flattening the perspective, removing any depth that the image might have had, and thus the use of close mikes must be handled with subtlety. David Griesinger has suggested that the use of stereo pairs of mikes as spots can help enormously in removing this flattening effect, because the spill that results between spots is now in stereo rather than in mono and is perceived as reflections separated spatially from the main signal.

The recent development of cheaper digital signal processing (DSP) has made possible the use of delay lines, sometimes as an integral feature of digital mixer channels, to adjust the relative timing of spot mikes in relation to the main pair so as to prevent the distortion of distance, and to equalise the arrival times of distant mikes so that they do not exert a precedence 'pull' over the output of the main pair. Wöhr *et al.* described a detailed approach to this, termed 'room-related balancing', in which the delays and levels of microphone signals are controlled so as to emulate the reflection characteristics of natural spaces (Wöhr *et al.* 1991). It is also possible to process the outputs of multiple mono sources to simulate binaural delays and head-related effects in order to create the effect of sounds at any position around the head when the result is monitored on headphones, as described in Chapter 3.

6.4 Three-channel techniques

The extension of two-channel recording techniques to three front channels is a useful step if considering how to derive signals for the front channels of a surround sound mix. This takes us back

Figure 6.19 (a) Bell Labs' original three-channel panning law (after Snow, 1953, and Gerzon, 1990). (The original Snow diagram shows 'source position' in feet from centre as well as panpot setting, because his experiments related to simulating the locations of real sources placed in front of three spaced microphones using panning. This is omitted here because the distances are arbitrary and might be confusing. It is not entirely clear what angle was subtended by the loudspeakers here, but from other evidence it appears to be about ±35°. In any case, the Bell Labs' law was intended for large auditorium listening where listener position could not easily be controlled.) (b) Gerzon's modified 'psychoacoustic' proposal for three-channel panning. Notice the negative (anti-phase) gain settings at some positions. (This is one of a family of such panning laws, in this case optimised for low frequency imaging with loudspeakers at ±45°.)

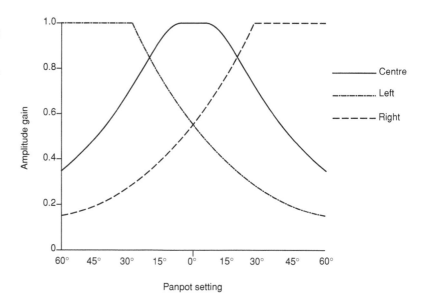

(a)

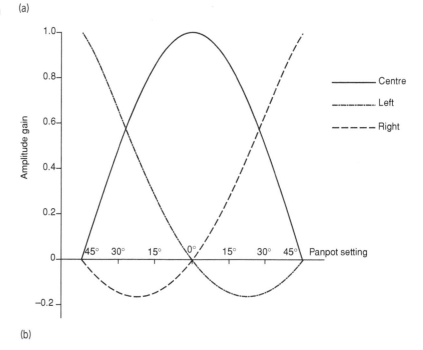

(b)

to some of the original proposals for three-channel stereo from the 1930s, although a number of the ideas for coincident arrays were not proposed at that time. The concept is expanded further in Chapter 7.

6.4.1 Three-channel panning laws

The presence of a centre speaker changes the panning laws and interchannel signal relationships required for accurate phantom imaging, as Michael Gerzon realised (see Section 4.1). Gerzon compared the original Bell Labs' three-channel panning law (see Figure 6.19(a)) with his own proposal for a psychoacoustic panning law that would optimise the interaural phase for localisation at low frequencies (see Figure 6.19(b)). (In his examples he assumed a 90° front sound stage, rather than the now more common 60°.) Gerzon's law contains some antiphase components, so signals panned right of centre give rise to a small negative component in the left speaker, and vice versa. The Bell law uses all positive components. He shows how the Bell law gives rise to reasonable HF localisation, but that this is not the same as LF localisation and can be unstable, especially for off-centre listeners (Figure 6.20). Adherence to a stable HF law, he asserts, is most appropriate for large auditorium listening as it is virtually impossible to optimise the LF law for multiple listening positions in large spaces. He also shows how his own LF law performs rather badly at HF, having a strong pull towards the

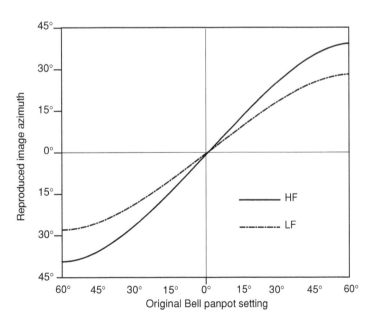

Figure 6.20 Gerzon's predictions of LF and HF localisation accuracy with Bell Labs' panning law.

centre at HF, and concludes that an optimal panning law should be both frequency and application dependent. He therefore proposes the use of a basic frequency-independent panpot such as that shown in Figure 6.19(b), coupled with the use of suitable 3-in/3-out psychoacoustic decoders, to deal with the signal on replay for the appropriate context.

The above is a typical Gerzon-esque conclusion – conceptually thorough and psychoacoustically optimal, but not particularly straightforward in practice. It relies on end users appreciating the importance of correct psychoacoustic decoding, and it breaks from the concept of pair-wise amplitude panning used on most mixing consoles ('film-style' panning). Typical three-channel (or more) panpots are rather crude devices, tending to work on simple positive-valued gain relationships between pairs of channels at a time, treating each pair of speakers (left-centre or centre-right) as a straight amplitude-panned pair as in two-channel stereo, using a law similar to that shown in Figure 6.21.

6.4.2 Three-channel microphone techniques

Spaced microphone techniques based on omnidirectional microphones can be moderately easily extended to three channels, although strict theoretical analysis is rather more complex because less work has been done on multiple source precedence effect in sound reproduction. If one treats the channels as individual sources then one can extend the precedence effect

Figure 6.21 A typical conventional three-channel panpot law using 'pairwise' constant power relationships.

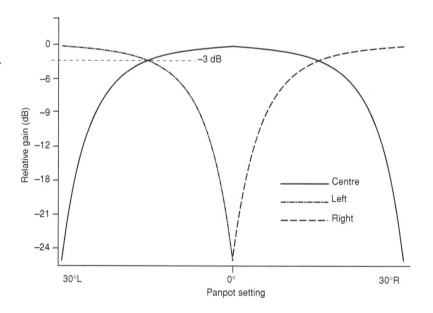

principle stated earlier to three sources emitting the same signal with different time delays, the image being localised towards the earliest one. This is covered further in Chapter 7, as it relates to surround microphone arrays, but it is noted that most theoretical experiments on the effects of time delays between multiple channels have tended to treat the microphones/loudspeakers in pairs, whereas in multi-microphone arrays a source in a particular position will give rise to signals from all channels with different time delays. Some have claimed that this will give rise to conflicting phantom images between pairs of microphones (e.g. one between centre and left, another between centre and right and yet another between left and right). This is unlikely to be the case in practice, as the precedence effect takes over and the hearing process tends to fuse the images and assign it one direction depending on the time delays and level differences involved. The focus or spread of the image may vary depending on the outputs from different loudspeakers, though.

Informal experiments by the author and others suggest that a convincing front image can be created by using an array similar to that of the Decca Tree described above, but this time with the centre microphone routed to the centre speaker, rather than being bridged between left and right. The spacing between the left and the right microphones may need to be slightly different than for the Decca Tree to give adequate sound stage width, and the dimensions need to be adjusted to suit the width of the sound stage and the distance of the mikes from the sources. (Griesinger has claimed that front imaging techniques based primarily on time delay are less tolerant of different listener positions than those based on amplitude differences or panning, because the image positions move too much as the listener position changes. A pleasing spatial effect is created, though, across a range of listening positions, even if the precision of the phantom images is not as great as that obtained with amplitude-based images.)

Coincident microphone techniques for three-channel front reproduction are somewhat more problematic than spaced techniques because the polar patterns required to arrive at something approaching the correct signal relationships at the listener's ears for optimal source localisation are often non-standard. McKinnie and Rumsey (1997) showed that it was possible to find some coincident three-channel arrangements using conventional polar patterns that gave rise to psychoacoustic parameters fulfilling some of Gerzon's criteria for optimal stereophonic imaging across the majority of the recording angle, and sometimes slightly beyond. Some of these are shown in Figure 6.22. The parameter r_e is used to denote the magnitude of the energy

Figure 6.22 Some proposed coincident arrays for three-channel stereo that use commonly available polar patterns and maintain Gerzon's r(e) factor close to 1 for optimum image stability (McKinnie and Rumsey, 1997).

All of these arrays use three microphones: an MS pair used to derive the left and right channel polar patterns (whose S microphone is always figure-eight and whose MS ratio is 0.377) and a separate microphone to feed the centre channel. Each array is shown with its corresponding gain plot showing relative amplitudes of left, centre and right outputs at different source angles. In these cases, supercardioid is a polar pattern between cardioid and hypercardioid. (a) Centre channel = figure-eight; M channel = supercardioid. This gives rise to strongest rear pickup of the three shown.

(a) Centre channel

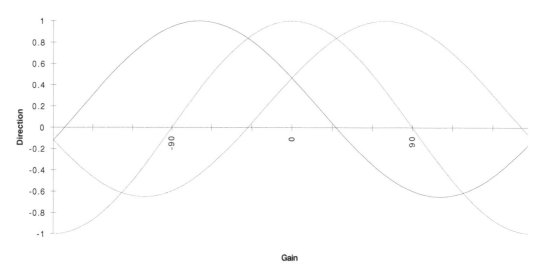

(a) M channel

(b) Centre channel =
hypercardioid; M channel =
supercardioid. This gives rise
to the lowest rear pickup of
the three shown.

(b) Centre channel

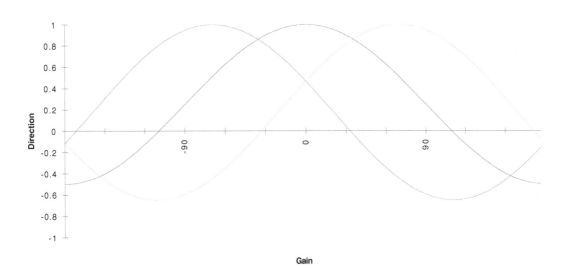

(b) M channel

continued

Figure 6.22 *continued*
(c) Centre channel =
supercardioid; M channel =
omni. This gives rise to an
intermediate level of rear
pickup.

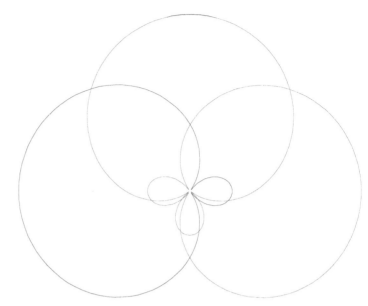

(c) Centre channel

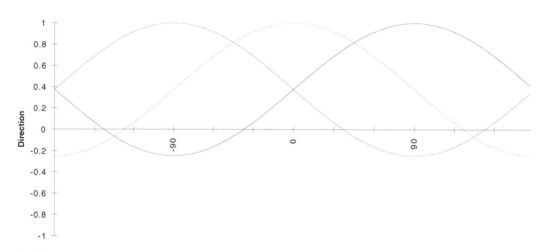

(c) M channel

vector that governs stability of phantom image localisation (primarily an HF phenomenon), and predicts the stability of phantom sources for off-centre listeners. Some arrays are better than others in this respect, and all show a strong tendency towards rear sound pickup, which may be problematic in some recording situations. The successful polar patterns are varying degrees of hypercardioid as a rule.

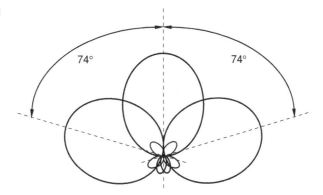

Figure 6.23 A three-channel coincident array using 2nd order microphones (after Cohen and Eargle, 1995).

Cohen and Eargle (1995) raise the oft-mentioned need for second-order directional microphones to solve the problem of coincident microphones for three-channel recording. Second order microphones have a narrower front pickup pattern but are very difficult to engineer and are not at all common today. A pattern described by $(0.5 + 0.5\cos\vartheta)\cos\vartheta$ is proposed as being quite suitable, being 3 dB down at about 37° from the main axis. This could be used to give good coverage in an array such as the one shown in Figure 6.23. They also suggest the possibility of using a conventional two-channel coincident pair with spaced outriggers (a typical classical recording arrangement), panning the coincident pair somewhere half left and right, and the outriggers fully left and right, based on experimentation.

6.4.3 'Stereo plus C'

Some recording engineers have coined the term 'stereo plus C' to refer to a three-channel front recording technique that is essentially the same as a two-channel technique but with the addition of a low level centre component from a separate microphone to solidify the centre image and to produce a signal from the centre loudspeaker. Such techniques have the advantage that they are sometimes easier to use in cases where two- and five-channel recordings are being made of the same event from a common set of microphones, and where a high level centre component would compromise a two-channel downmix.

References

Berkhout, A.J., de Vries, D. and Vogel, P. (1992). Wave front synthesis: a new direction in electroacoustics. Presented at *93rd AES Convention, San Francisco*, Preprint 3379.

Cohen, E. and Eargle, J. (1995). Audio in a 5.1 channel environment. Presented at *AES 99th Convention, New York*. Preprint 4071. Audio Engineering Society.

Cross, L. (1985). Performance assessment of studio microphones. *Recording Engineer and Producer*, February.

Dooley, W. and Streicher, R. (1982). MS stereo: a powerful technique for working in stereo. *J. Audio Eng. Soc.*, 30, 10, pp. 707–718.

Gayford, M. (ed.) (1994). *Microphone Engineering Handbook*. Focal Press, Oxford and Boston.

Gerzon, M. (1986). Stereo shuffling: new approach, old technique. *Studio Sound*, July.

Gerzon, M. (1990). Three channels: the future of stereo? *Studio Sound*, June, pp. 112–25.

Grignon, L. (1949). Experiments in stereophonic sound. *J. SMPTE*, 52, p. 280.

Hibbing (1989). XY and MS microphone techniques in comparison. Presented at *86th AES Convention, Hamburg*. Preprint 2811. Audio Engineering Society.

Lipshitz, S. (1986). Stereo microphone techniques: are the purists wrong? *J. Audio Eng. Soc.*, 34, 9, pp. 716–735.

McKinnie, D. and Rumsey, F. (1997) Coincident microphone techniques for three channel stereophonic reproduction. Presented at *102nd AES Convention, Munich*. Preprint 4429. Audio Engineering Society.

Snow, W. (1953) Basic principles of stereophonic sound. *JSMPTE*, 61, pp. 567–589.

Theile, G. (2000) Multichannel natural music recording based on psychoacoustic principles. Presented at *108th AES Convention, Paris, 19–22 February*. Preprint 5156. Audio Engineering Society.

Williams, M. (1987) Unified theory of microphone systems for stereophonic sound recording. Presented at *82nd AES Convention, London*. Preprint 2466. Audio Engineering Society.

Wöhr, M. *et al.* (1991) Room-related balancing technique: a method for optimizing recording quality. *J. Audio Eng. Soc.*, 39, 9, pp. 623–631.

7 Surround sound recording techniques

7.1 Surround sound microphone technique

This chapter deals with the extension of conventional two-channel recording technique to multiple channels for surround sound applications, concentrating on standard 5(.1)-channel reproduction. Many of the concepts described here have at least some basis in conventional two-channel stereo, although analysis of the psychoacoustics of 5.1 surround has been nothing like as exhaustively investigated to date. Consequently a number of the techniques described below are at a relatively early stage of development and are still being evaluated.

The chapter begins with a review of microphone techniques that have been proposed for the pickup of natural acoustic sources in surround, followed by a discussion of multichannel panning and mixing techniques, mixing aesthetics and artificial reverberation, for use with more artificial forms of production such as pop music. Film sound approaches are not covered in any detail as they are well established and not the main theme of this book. The chapter concludes with a section on conversion between surround formats and two-channel stereo, as well as vice versa.

7.1.1 Principles of surround sound microphone technique

Surround sound microphone technique, as discussed here, is unashamedly biased towards the pickup of sound for 5.1 surround, although Ambisonic techniques are also covered because they are well documented and can be reproduced over 5-channel loudspeaker systems if required, using suitable decoders. The techniques described in this section are most appropriate for use when the spatial acoustics of the environment are as important as those of the sources within, such as in classical music and other 'natural' recording. These microphone techniques tend to split into two main groups: those that are based on a single array of microphones in reasonably close proximity to each other, and those that treat the front and rear channels separately. The former are usually based on some theory that attempts to generate phantom images with different degrees of accuracy around the full 360° in the horizontal plane. The latter usually have a front array providing reasonably accurate phantom images in the front, coupled with a separate means of capturing the ambient sound of the recording space (often for feeding to all channels in varying degrees). It is rare for such microphone techniques to provide a separate feed for the LFE channel, so they are really 5-channel techniques not 5.1-channel techniques.

The concept of a 'main array' or 'main microphone configuration' for stereo sound recording is unusual to some recording engineers, possibly being a more European than American concept. The traditional European approach has tended to involve starting with a main microphone technique of some sort that provides a basic stereo image and captures the spatial effect of the recording environment in an aesthetically satisfactory way, and then supporting this subtly to varying degrees with spot mikes as necessary. It has been suggested by some that many balances in fact end up with more sound coming from the spot mikes than from the main array in practice, and that in this case it is the spatial treatment of the spot mikes and any artificial reverberation that will have most effect on the perceived result. This is covered in the next section and the issue is open to users for further experimentation.

Those of a sceptical persuasion can cite numerous research papers that show how difficult it is to create stable phantom images to the sides of a listener in a standard 5.1 surround configuration, using simple pairwise amplitude or time differences. These problems were summarised in Section 2.1.4. To recap, simple amplitude or time differences between side pairs of loudspeakers such as L and LS or R and RS are incapable of generating suitable differences between the ears of a front-facing listener to create stable images, although it has been found that amplitude differences give slightly more stable results than time differences. If the listener turns to face the speaker pair then the situation may be improved somewhat, but the subtended angle of about 80° still results in something of a hole in the middle and the same problem as before then applies to the front and rear pairs. Phantom sources can be created between the rear speakers but the angle is again quite great (about 140°), leading to a potential hole in the middle for many techniques, with the sound pulling towards the loudspeakers. This suggests a gloomy prognosis for those techniques attempting to provide 360° phantom imaging, and might suggest that one would be better off working with 2- or 3-channel stereo in the front and decorrelated ambient signals in the rear.

A number of factors, though, may make the prognosis for 360° imaging less gloomy than the previous paragraph would suggest. Firstly, Ambisonic panning and microphone techniques (see below) can be used to generate appropriate signals for reasonable 360° imaging, but generally only for a limited range of listening positions and best with a square or rectangular speaker array. Using a 5.1 loudspeaker array makes it difficult to get such good results at the sides and the rear, but gives enhanced image stability at the front. Secondly, 5-channel microphone arrays produce *some* output to all five channels, no matter where the source, so it is not simply differences between loudspeaker pairs that one must consider but differences between signals from all five loudspeakers. The effect of different time delays and levels from all the possible combinations of channels has not yet been fully explored, and the subjective results from arrays that are based on attempts at 360° imaging are often quite convincing, suggesting that one should investigate further before writing them off as unworkable.

All this said, there is no escaping the fact that it is easiest to create images where there are loudspeakers, and that phantom images between loudspeakers subtending wide angles tend to be unstable or 'hole-in-the-middly'. Given this unavoidable aspect of surround sound psychoacoustics, one should always expect imaging in standard 5-channel replay systems to be best between the front loudspeakers, only moderate to the rear, and highly variable to the sides (see Figure 7.1). Since the majority of material one listens to tends to conforms to this paradigm in any case (primary sources in front, secondary content to the sides and rear), the problem is possibly not as serious as it might seem. It is tempting to rail against the 5.1 standards for limiting the capacity of sound reproduction to achieve good 360° localisation, and consequently to promote the use of alternative loudspeaker arrangements that might be more appropriate, but the 5.1 standard is a compromise that has taken years to come about and represents the best chance we have for the time being of

Figure 7.1 Imaging accuracy in five-channel surround sound reproduction.

Good phantom images between left, centre and right loudspeakers

Typically poor and unstable phantom images between front and surround loudspeakers

Typically poor and unstable phantom images between front and surround loudspeakers

Only moderately satisfactory phantom images between rear loudspeakers, with a tendency towards a 'hole in the middle'

improving upon the spatial experience offered by two-channel stereo, at least for the mass market.

One must accept also that the majority of consumer systems will have great variability in the location and nature of the surround loudspeakers, making it unwise to set too much store by the ability of such systems to enable accurate sound field reconstruction in the home. Better, it seems, would be to acknowledge the limitations of such systems and to create recordings that work best on a properly configured reproduction arrangement but do not rely on 100% adherence to a particular reproduction alignment and layout, or on a limited 'hot spot' listening position. Surround sound provides an opportunity to create something that works over a much wider range of listening positions than two-channel stereo, does not collapse rapidly into the nearest loudspeaker when one moves, and enhances the spatial listening experience.

7.1.2 Five-channel 'main microphone' arrays

Recent interest in 5-channel recording has spawned a number of variants on a common theme involving fairly closely spaced microphones (often cardioids) configured in a 5-point array. The basis of most of these arrays is pair-wise time-intensity trading,

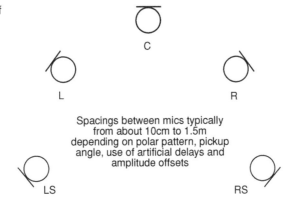

Figure 7.2 Generic layout of five-channel microphone arrays based on time–amplitude trading.

C

L   R

Spacings between mics typically
from about 10cm to 1.5m
depending on polar pattern, pickup
angle, use of artificial delays and
amplitude offsets

LS   RS

using techniques similar to those described in Section 6.2.4, usually treating adjacent microphones as pairs covering a particular section of the recording angle around the array and possibly hoping that the signals from the other microphones will be either low enough in level or long enough delayed not to affect the image in the sector concerned too much. The generic layout of such arrays is shown in Figure 7.2. Cardioids tend to be favoured because of the increased direct-to-reverberant pickup they offer, and the interchannel level differences created for relatively modest spacings and angles, enabling the array to be mounted on a single piece of metalwork. The centre microphone is typically spaced slightly forward of the L and R microphones thereby introducing a useful time advance in the centre channel for centre front sources.

The spacing and angles between the capsules are typically based on the so-called 'Williams curves' mentioned in Chapter 6, based on time and amplitude differences required between single pairs of microphones to create phantom sources in particular locations. (NB: the Williams curves were based on two-channel pairs and loudspeaker reproduction in front of the listener, and it is not necessarily the case that the same technique can be applied to create images between pairs at the sides of the listener, nor that the same level and time differences will be suitable. There is some evidence that different delays are needed between side and rear pairs than those used between front pairs. The effects of inter-pair crosstalk also need to be studied further, as discussed below.)

Some of the techniques based on this general principle have been given proprietary names by the people who have developed or marketed them as commercial products. A detailed theoretical treatment of such an array has been provided by Williams and

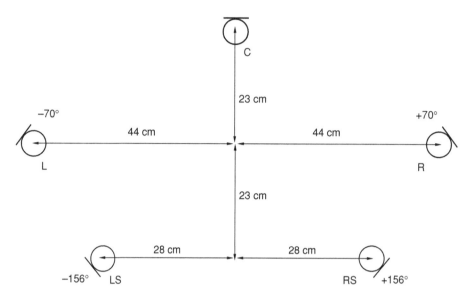

Figure 7.3 Five-channel microphone array using cardioids, one of a family of arrays designed by Williams and Le Dû. In this example the front triplet should be attenuated 2.4 dB with respect to the rear pair.

Le Dû (1999, 2000), and commercial systems using similar principles have been labelled TSRS (for True Space Recording System) by Mora and Jacques (1998).

Williams and Le Dû describe a process they call 'critical linking' between the recording angles or sectors covered by each pair of microphones in an array. In order that each pair of microphones covers a sector that does not overlap with any other, certain time and gain offsets are introduced between channels, either using positional alterations of the microphones or electronic offsets, or a combination of the two. One possible configuration of many from such an optimisation process is pictured in Figure 7.3. To satisfy the critical linking requirements for this particular array the front triplet is attenuated by 2.4 dB in relation to the back pair. The authors have written software to assist in the somewhat long-winded process of optimising their arrays, and interested readers are referred to their papers for further details.

Mora and Jacques, in describing their array of a similar nature (called TSRS for 'true space recording system'), show experimental data comparing human localisation performance in respect of noise bursts in different locations around the listener. First they measured localisation accuracy using real sources (loudspeakers) to generate the noise bursts and then they repeated the experiment using a TSRS array recording of the same sources reproduced over a 5-channel loudspeaker system arranged according to ITU BS.775 specifications. As expected, localisation errors at the sides are much more common than at the front and back, but they compare this with similarly erroneous judgements noticed for the real sources in those places, so they claim the results of their microphone array are not too dissimilar to those that would arise in natural listening anyway.

The issue of crosstalk between the pairs covering the recording 'sectors' (e.g. L-C or C-R) has been raised by some, including Theile (2000) who asserts that crosstalk from other microphones in the array (other than the pair in question) should be reduced as far as possible otherwise it will blur the image and introduce colouration. Theile also suggests that multiple phantom sources will be created, arising from the signal differences between the various pairs involved in a three-microphone front array, and that these will give rise to multiple comb filtering effects when combined to form a two-channel downmix. He suggests that the issue of crosstalk is important because it is only 1–2 ms delayed and the channel separation is generally less than 6 dB. Parts of this argument may be questioned, though, as it is more likely that a single 'fused' phantom source will be perceived whose size, stability and position are governed by the relevant amplitude and time differences between the signals. Predicting where it will be and what the effects of the multiple signals will be is the most complicated factor, requiring further subjective tests. Williams has attempted to show in subsequent papers that the effects of crosstalk are minimal at most angles, but the theoretical basis for this claim is not entirely clear. While the levels and time delays of the crosstalk may be outside the windows tested by Simonsen in his original experiments on time–amplitude trading in microphone arrays, more conclusive evidence is needed that the crosstalk does not matter. (The crosstalk from other microphones could possibly be beneficial in creating lateral images owing to precedence effects between microphones on opposite *sides* of the array, rather than between adjacent mikes.)

Some success has also been had by the author's colleagues using omni microphones instead of cardioids, with appropriate adjustments to the spacings according to 'Williams-style' time-amplitude trading curves (also with modifications to correct for different inter-loudspeaker angles and spacings to the sides and rear), and these tend to give better overall sound quality but (possibly unsurprisingly) poorer front imaging. Side imaging has proved to be better than expected with omni arrays.

The closeness between the microphones in these arrays is likely to result in only modest low frequency decorrelation between the channels. If, as Griesinger suggests, good LF decorrelation is

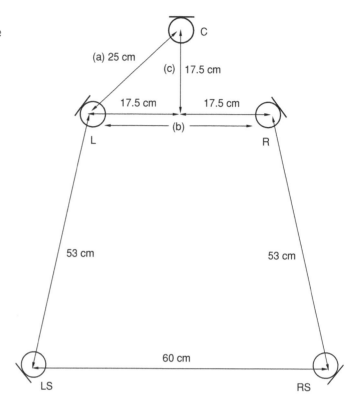

important for creating a sense of spaciousness, these 'near-coincident' or 'semi-correlated' techniques will be less spacious than more widely spaced microphone arrays. Furthermore, the strong dependence of these arrays on precedence effect cues for localisation makes their performance quite dependent on listener position and front–rear balance.

The INA (Ideale Nieren Anordnung) or 'Ideal Cardioid Array' is described by Hermann and Henkels (1998) as a three-channel front array of cardioids (INA-3) coupled with two surround microphones of the same polar pattern (making it into an INA-5 array). One configuration of this is shown in Figure 7.4, and a commercial implementation by Brauner is pictured in Figure 7.5. (It is assumed that the 5-channel array is intended to create images throughout 360°, in a similar way to Williams *et al.*, but the reasons for the spacing of the rear microphones are not entirely clear.) Table 7.1 shows some possible combinations of microphone spacing and recording angle for the front three microphones of this proposed array. In the commercial implementation the capsules can be moved and rotated and their polar patterns can be varied. The configuration shown in Figure 7.4 is

Figure 7.5 SPL/Brauner Atmos 5.1/ASM5 system.

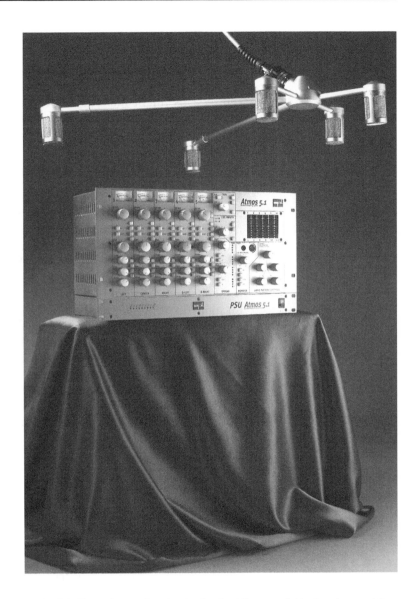

Table 7.1 Dimensions and angles for the front three cardioid microphones of the INA array (see Figure 7.4). Note that the angle between the outer microphones should be the same as the recording angle.

Recording angle (ϑ)°	Microphone spacing (a) cm	Microphone spacing (b) cm	Array depth (c) cm
100	69	126	29
120	53	92	27
140	41	68	24
160	32	49	21
180	25	35	17.5

termed an 'Atmokreuz' (atmosphere cross) by the authors. Its large front recording angle of 180° means that to use it as a main microphone it would have to be placed very close to the source unless all the sources were to appear to come from near the centre. This might make it less well placed for the surrounds. Such a configuration may be more suitable for general pickup slightly further back in the hall.

7.1.3 Separate treatment of front imaging and ambience

Many alternative approaches to basic microphone coverage for 5.1 surround treat the stereo imaging of front signals separately from the capture of a natural-sounding spatial reverberation and reflection component, and some are hybrid approaches without a clear theoretical basis. Most do this by adopting a three-channel variant on a conventional two-channel technique for the front channels, as introduced in Chapter 6 (sometimes optimised for more direct sound than in a two-channel array), coupled with a more or less decorrelated combination of microphones in a different location for capturing spatial ambience (sometimes fed just to the surrounds, other times to both front and surrounds). Sometimes the front microphones also contribute to the capture of spatial ambience, depending on the proportion of direct to reflected sound picked up, but the essential point here is that the front and rear microphones are not intentionally configured as an attempt at a 360° imaging array.

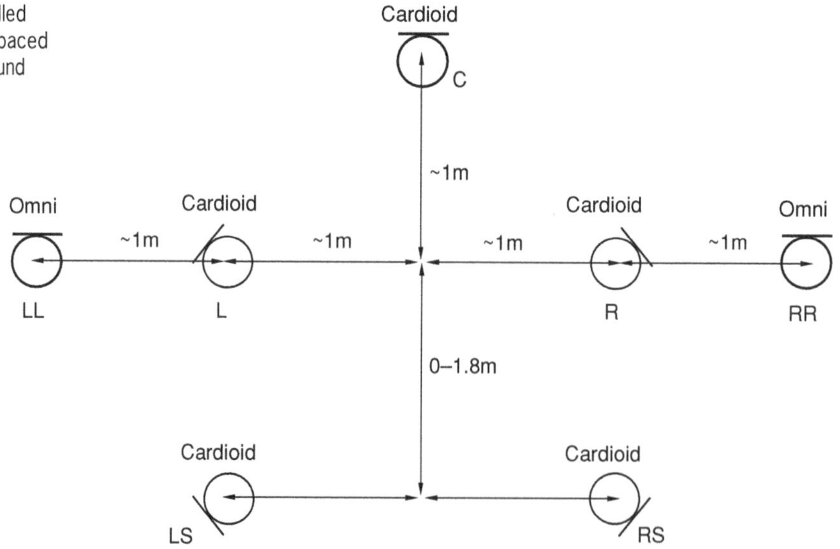

Figure 7.6 The so-called 'Fukada Tree' of five spaced microphones for surround recording.

Fukada *et al.* (1997) describe techniques for recording ambient sound separately from front signals. The so-called 'Fukada Tree', shown in Figure 7.6, is based on a Decca Tree, but instead of using omni mikes it mainly uses cardioids. The reason for this is to reduce the amount of reverberant sound pickup by the front mikes. Omni outriggers are sometimes added as shown, typically panned between L–LS and R–RS, in an attempt to increase the breadth of orchestral pickup and to integrate front and rear elements. The rear mikes are also cardioids and are typically located at approximately the critical distance of the space concerned (where the direct and reverberant components are equal). They are sometimes spaced further back than the front mikes by nearly 2 metres, although the dimensions of the tree can be varied according to the situation, distance, etc. (Variants are known that have the rear mikes quite close to the front ones, for example.) The spacing between the mikes more closely fulfils Griesinger's requirements for the decorrelated microphone signals needed to create spaciousness, depending on the critical distance of the space in which they are used. Mikes should be separated by at least the room's critical distance for adequate decorrelation. The front imaging of such an array would be similar to that of an ordinary Decca Tree (not bad, but not as precise as some other techniques), but the frequency response advantages of omnis would be lost.

Erdo Groot of Polyhymnia International has developed a variant on this approach that uses omnis instead of cardioids, to take advantage of their better sound quality. Using an array of omnis separated by about 3 metres between left–right and front–back, he achieves a spacious result where the rear channels are well integrated with the front. The centre mike is placed slightly forward of left and right. It is claimed that placing the rear omnis too far away from the front tree makes the rear sound detached from the front image, so one gets a distinct echo or repeat of the front sound from the rear.

Hamasaki of NHK has proposed an arrangement based on near-coincident cardioids (30 cm) separated by a baffle, as shown in Figure 7.7. Here the centre cardioid is placed slightly forward of left and right, and omni outriggers are spaced by about 3 m. These omnis are low-pass filtered at 250 Hz and mixed with the left and right front signals to improve the LF sound quality. Left and right surround cardioids are spaced about 2–3 m behind the front cardioids and 3 m apart. An ambience array is used further back, consisting of four figure-eight mikes facing sideways, spaced by about 1 m, to capture lateral reflections, fed to the four outer channels. This is placed high in the recording space.

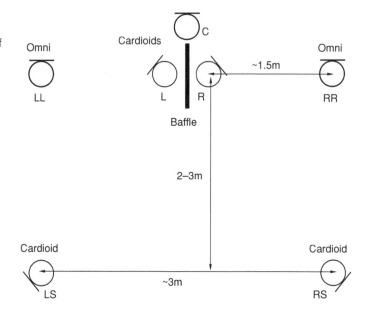

Figure 7.7 A surround technique proposed by Hamasaki (NHK) consisting of a cardioid array, omni outriggers and separate ambience matrix.

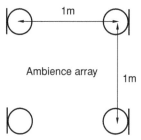

Mason and Rumsey (1999) described a subjective comparison between different approaches to ambient sound recording for the surround channels. Comparing delayed and undelayed rear-facing cardioids, similar to the Fukada arrangement, with distant omni microphones, they found that delayed rear-facing cardioids were preferred to the other techniques for spatial impression and front imaging results. The delay was of the order of 30 ms to provide a precedence effect 'pull' towards the front

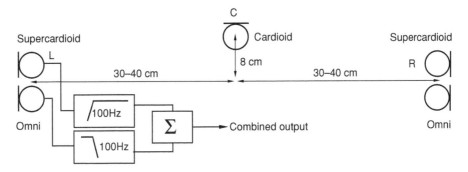

Figure 7.8 Theile's proposed three-channel array for front pickup using supercardioids for the outer mikes, crossed over to omni at LF. The spacing depends on the recording angle (C–R = 40 cm for 90° and 30 cm for 110°).

and avoid direct sound from the front 'bleeding' into the surrounds (rather as with Dolby Surround).

Theile proposes a front microphone arrangement shown in Figure 7.8. While superficially similar to the front arrays described in Section 7.1.2, he reduces crosstalk between the channels by the use of supercardioid microphones at ±90° for the left and right channels and a cardioid for the centre. (Supercardioids are more directional than cardioids and have the highest direct/reverberant pickup ratio of any first-order directional microphone. They have a smaller rear lobe than hypercardioids.) Theile's rationale behind this proposal is the avoidance of crosstalk between the front segments. He proposes to enhance the LF response of the array by using a hybrid microphone for left and right, that crosses over to omni below 100 Hz, thereby restoring the otherwise poor LF response. The centre channel is high pass filtered above 100 Hz. Furthermore, the response of the supercardioids should be equalised to have a flat response to signals at about 30° to the front of the array (they would normally sound quite coloured at this angle). Such a proposal demands some fairly complex microphone mounting, and possibly the development of a hybrid capsule with appropriate crossover and equalisation. A home-made version could also possibly be constructed. Schoeps has developed a prototype of this array, and it has been christened 'OCT' for 'Optimum Cardioid Triangle'.

For the ambient sound signal, Theile proposes the use of a crossed configuration of microphones, that has been christened the 'IRT cross' or 'atmo-cross'. This is shown in Figure 7.9. The microphones are either cardioids or omnis, and the spacing is chosen according to the degree of correlation desired between the channels. Theile suggests 25 cm for cardioids and about 40 cm for omnis, but says that this is open to experimentation. Small spacings are appropriate for more accurate imaging of

Figure 7.9 The IRT 'atmo-cross' designed for picking up ambient sound for routing to four loudspeaker channels (omitting the centre). Mikes can be cardioids or omnis (wider spacing for omnis).

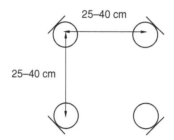

reflection sources at the hot spot, whereas larger spacings are appropriate for providing diffuse reverberation over a large listening area. The signals are mixed in to L, R, LS and RS channels, but not the centre.

Klepko (1997) proposed a front array with a cardioid in the centre and supercardioids for L and R, arranged as shown in Figure 7.10. The angle of the supercardioids is much smaller than Theile's, the spacing is ±17.5 cm and the microphones are in line. The theoretical basis for this configuration is not fully

Figure 7.10 Klepko's proposal for microphone pickup in a five-channel system. The centre front mike is a cardioid whereas the outer mikes are supercardioid. A dummy head is used for the rear channels.

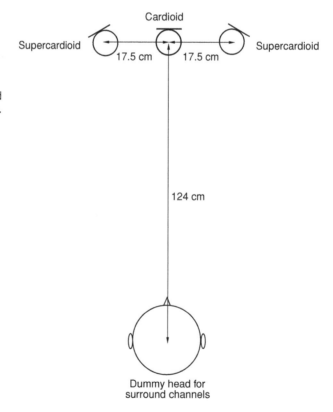

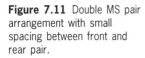

Figure 7.11 Double MS pair arrangement with small spacing between front and rear pair.

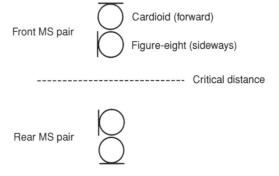

explained and results in a large overlap of the recording angle of the two pairs (L-C and C-R) in the centre. Klepko also proposes the use of a pseudo-binaural technique for the rear channels, as described in Section 7.1.4.

A 'double MS' technique has been proposed by Curt Wittig and others, shown in Figure 7.11. Two MS pairs (see Chapter 6) are used, one for the front channels and one for the rear. The centre channel can be fed from the front M microphone. The rear pair is placed at or just beyond the room's critical distance. S gain can be varied to alter the image width in either sector, and the M mike's polar pattern can be chosen for the desired directional response (it would typically be a cardioid). Others have suggested using a fifth microphone (a cardioid) in front of the forward MS pair, to feed the centre channel, delayed to time align it with the pair. If the front and rear MS pairs are co-located it may be necessary to delay the rear channels somewhat (10–30 ms) so as to reduce perceived spill from front sources into rear channels. In a co-located situation the same figure-eight microphone could be used as the S channel for both front and back pairs.

In general, the signals from separate ambience microphones fed to the rear loudspeakers may often be made less obtrusive and front-back 'spill' may be reduced by rolling off the high frequency content of the rear channels. Some additional delay may also assist in the process of integrating the rear channel ambience. The precise values of delay and equalisation can only really be arrived at by experimentation in each situation.

7.1.4 Pseudo-binaural techniques

As with two-channel stereo, some engineers have experimented with pseudo-binaural recording techniques intended for loudspeaker reproduction. Jerry Bruck adapted the Schoeps

Figure 7.12 (a) Schoeps
KFM360 sphere microphone
with additional figure-eights
near the surface mounted
omnis. (b) KFM360 control
box. (Courtesy of
Schalltechnik Dr.-Ing. Schoeps
GmbH).

(a)

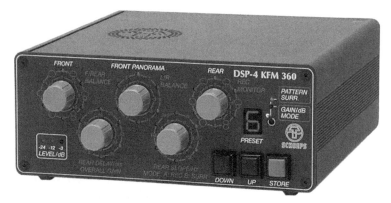

(b)

'Sphere' microphone, described in Section 6.2.5, for surround sound purposes by adding bi-directional (figure-eight) microphones near to the 'ears' (omni mikes) of the sphere, with their main axis front–back, as pictured in Figure 7.12 (Bruck, 1997). This microphone is now manufactured by Schoeps as the KFM360. The figure-eights are mounted just below the sphere transducers so as to affect their frequency response in as benign a way as possible for horizontal sources. The outputs from the figure-eight and the omni at each side of the sphere are MS matrixed to create pairs of roughly back-to-back cardioids facing sideways. The size of the sphere creates an approximately ORTF spacing between the side-facing pairs. The matrixed output of this microphone can be used to feed four of the channels in a 5-channel reproduction format (L, R, LS and RS). A Schoeps processing unit can be used to derive an equalised centre channel from the front two, and enables the patterns of front and rear coverage to be modified.

Michael Bishop of Telarc has reportedly adapted the 'double MS' technique described in Section 7.1.3 by using MS pairs facing sideways, and a dummy head some 1–2.5 m in front, as shown in Figure 7.13 (Mitchell, 1999). The MS pairs are used between side pairs of channels (L and LS, R and RS) and line-up is apparently tricky. The dummy head is a model equalised for a natural response on loudspeakers (Neumann KU100) and is used for the front image.

Klepko proposed using a binaural technique for the rear channels of a surround recording array, the front part of which was described in Section 7.1.3. This is pictured in Figure 7.16. The reason for the 124 cm spacing between the front array and

Figure 7.13 Double MS pairs facing sideways used to feed the side pairs of channels combined with a dummy head facing forwards to feed the front image.

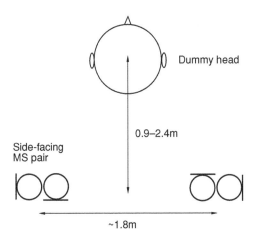

Dummy head

0.9–2.4m

Side-facing
MS pair

~1.8m

the dummy head is not explained, but it is probably intended to introduce a small delay and decorrelation effect. He claimed that the head's shadow would act to cancel high frequency interaural crosstalk to a degree during reproduction, and that this was best achieved with loudspeakers at ±90°. The dummy head was equalised to compensate for the most prominent spectral aberrations for that listening angle on loudspeakers.

7.1.5 Multimicrophone techniques

Most real recording involves the use of spot microphones in addition to a main microphone technique of some sort, indeed in many situations the spot microphones may end up at higher levels than the main microphone or there may be no main microphone. The principles outlined in Chapter 6 still apply in surround mixing, but now one has the issue of surround panning to contend with. The principles of this are covered in more detail in Section 7.2. Some aesthetic considerations relating to the panning of multiple sources are discussed in Section 7.3.

Some engineers report success with the use of multiple sphere microphones for surround balances, which is probably the result of the additional spatial cues generated by using a 'stereo' spot mike rather than a mono one, avoiding the flatness and lack of depth often associated with panned mono sources. Artificial reverberation of some sort is almost always helpful when trying to add spatial enhancement to panned mono sources, and some engineers prefer to use amplitude-panned signals to create a good balance in the front image, plus artificial reflections and reverberation to create a sense of spaciousness and depth.

7.1.6 Ambisonic or 'Soundfield' microphone principles

The so-called 'Soundfield' microphone, pictured in Figure 7.14, is designed for picking up full periphonic sound in the Ambisonic A-format (see Section 4.8.2), and is coupled with a control box designed for converting the microphone output into both the B-format and the D-format. Decoders can be created for using the output of the Soundfield microphone with a 5.1-channel loudspeaker array, including that recently introduced by Soundfield Research. The full periphonic effect can only be obtained by reproduction through a suitable periphonic decoder and the use of a tetrahedral loudspeaker array with a height component, but the effect is quite stunning and worth the effort.

The Soundfield microphone is capable of being steered electrically by using the control box, either in terms of azimuth, elevation, tilt

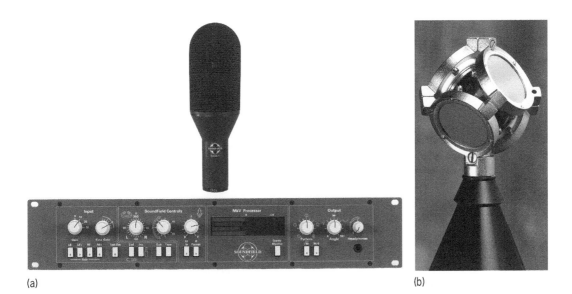

(a)

(b)

Figure 7.14 (a) The Soundfield microphone and accompanying control box, (b) capsule arrangement (Courtesy Soundfield Research).

or dominance, and as such it is also a particularly useful stereo microphone for two-channel work. The microphone encodes directional information in all planes, including the pressure and velocity components of indirect and reverberant sounds.

Figure 7.14(b) shows the physical capsule arrangement of the microphone, which were shown diagramatically in Figure 4.15. Four capsules with sub-cardioid polar patterns (between cardioid and omni, with a response equal to $2+\cos\vartheta$) are mounted so as to face in the A-format directions, with electronic equalisation to compensate for the inter-capsule spacing, such that the output of the microphone truly represents the sound-field at a point (true coincidence is maintained up to about 10 kHz). The capsules are matched very closely and each contributes an equal amount to the B-format signal, thus result-ing in cancellation between variations in inherent capsule responses. The A-format signal from the microphone can be converted to B-format according to the equations given in Section 4.8.2.

The combination of B-format signals in various proportions can be used to derive virtually any polar pattern in a coincident configuration, using a simple circuit as shown in Figure 7.15 (two-channel example). Crossed figure-eights are the most obvious and simple stereo pair to synthesise, since this requires the sum-and-difference of X and Y, whilst a pattern such as crossed cardioids requires that the omni component be used also, such that:

Figure 7.15 Circuit used for controlling stereo angle and polar pattern in Soundfield microphone. (Courtesy of Ken Farrar).

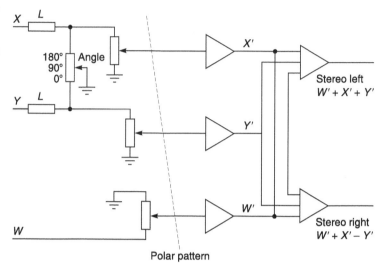

$$\text{Left} = W + (X/2) + (Y/2)$$

$$\text{Right} = W + (X/2) - (Y/2)$$

From the circuit it will be seen that a control also exists for adjusting the effective angle between the synthesised pair of microphones, and that this works by varying the ratio between X and Y in a sine/cosine relationship.

The microphone may be controlled, without physical re-orientation, so as to 'point' in virtually any direction (see Figure 7.16). It may also be electrically inverted, so that it may be used

Figure 7.16 Azimuth, elevation and dominance in Soundfield microphone.

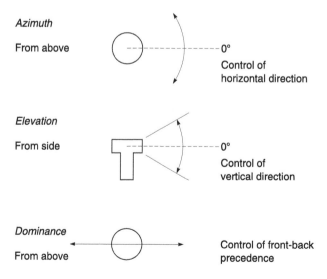

Figure 7.17 Circuit used for azimuth control. (Courtesy of Ken Farrar).

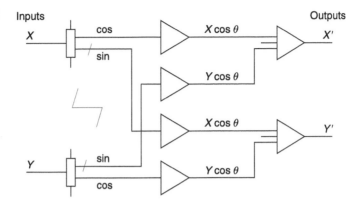

upside-down. Inversion of the microphone is made possible by providing a switch that reverses the phase of Y and Z components. W and X may remain unchanged since their directions do not change if the microphone is used upside-down. Azimuth is controlled by taking X and Y components and passing them through twin-ganged sine/cosine potentiometers, as shown in Figure 7.17, and processing them such that two new X and Y components are produced (X' and Y') which are respectively:

$$X' = X\cos\vartheta + Y\sin\vartheta$$

$$Y' = Y\cos\vartheta - X\sin\vartheta$$

Elevation (over a range of ±45°) is controlled by acting on X and Z to produce X' and Z', using the circuit shown in Figure 7.18.

Figure 7.18 Circuit used for elevation control. (Courtesy of Ken Farrar).

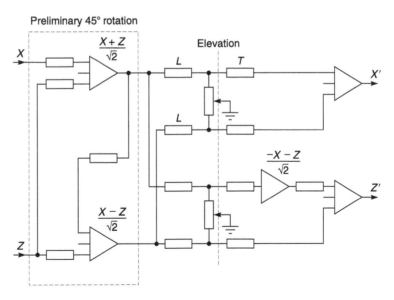

Figure 7.19 Circuit used for dominance control. (Courtesy of Ken Farrar).

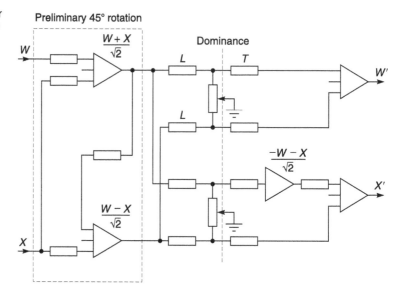

Firstly, a circuit produces sum and difference signals equivalent to rotations through 45° up and down in the vertical plane, and then proceeds to combine these rotated components in appropriate proportions corresponding to varying angles between ±45°. Dominance is controlled by varying the polar diagram of the W component, such that it ceases to be an omni, and becomes more cardioid, either favouring sounds from the front or from the rear. It has been described by the designer as a microphone 'zoom' control, and may be used to move the microphone 'closer' to a source by rejecting a greater proportion of rear pickup, or further away by making W more cardioid in the reverse direction. This is achieved by adding or subtracting amounts of X to and from W, using the circuit shown in Figure 7.19.

7.2 Multichannel panning techniques

The panning of signals between more than two loudspeakers presents a number of psychoacoustic problems, particularly with regard to appropriate energy distribution of signals, accuracy of phantom source localisation, off centre listening and sound timbre. A number of different solutions have been proposed, in addition to the relatively crude pairwise approach used in much film sound, and some of these are outlined below. The issue of source distance simulation is also discussed.

It is possibly relevant here to quote Michael Gerzon's criteria for a good panning law for surround sound (Gerzon, 1992b):

The aim of a good panpot law is to take monophonic sounds, and to give each one amplitude gains, one for each loudspeaker, dependent on the intended illusory directional localisation of that sound, such that the resulting reproduced sound provides a convincing and sharp phantom illusory image. Such a good panpot law should provide a smoothly continuous range of image directions for any direction between those of the two outermost loudspeakers, with no 'bunching' of images close to any one direction or 'holes' in which the illusory imaging is very poor.

7.2.1 Pairwise amplitude panning

Pairwise amplitude panning is the type of pan control most recording engineers are familiar with, as it is the approach used on most two-channel mixers. As described in Chapter 6, it involves adjusting the relative amplitudes between a pair of adjacent loudspeakers so as to create a phantom image at some point between them. This has been extended to three front channels as described earlier, and is also sometimes used for panning between side loudspeakers (e.g. L and LS) and rear loudspeakers. The typical sine/cosine panning law devised by Blumlein for two-channel stereo is often simply extended to more loudspeakers. Most such panners are constructed so as to ensure constant power as sources are panned to different combinations of loudspeakers, so that the approximate loudness of signals remains constant. As previous discussions have explained (see particularly Section 2.1.4), panning using amplitude or time differences between widely spaced side loudspeakers is not particularly successful at creating accurate phantom images. Side images tend not to move linearly as they are panned and tend to jump quickly from front to back. Spectral differences resulting from differing HRTFs of front and rear sound tend to result in sources appearing to be spectrally split or 'smeared' when panned to the sides.

As Holman describes (Holman, 1999), in some mixers designed for five-channel surround work, particularly in the film domain, separate panners are provided for L-C-R, LS-RS, and front-surround. Combinations of positions of these amplitude panners enables sounds to be moved to various locations, but some more successfully than others. For example sounds panned so that some energy is emanating from all loudspeakers (say, panned centrally on all three pots) tend to sound diffuse for centre listeners, and in the nearest loudspeaker for those sitting off centre.

Joystick panners combine these amplitude relationships under the control of a single lever that enables a sound to be 'placed' dynamically anywhere in the surround sound field, but it is certainly true that the moving effects made possible by these joysticks are often unconvincing and need to be used with experience and care.

Those that proposed the 3/2 standard for surround sound were well aware of this problem and it was one of the main reasons why the surround channels were proposed as ambience, 'room' or effect channels to accompany three-channel stereo at the front. Recent data from Martin *et al.* were mentioned in Chapter 2, indicating how uncertain the location of side images was when panned in a 3/2 layout. Here we can see additional data from Theile and Plenge (1977) demonstrating much the same thing (see Figure 7.20).

Research undertaken by Jim West at the University of Miami (West, 1999) showed that despite the limitations of constant power 'pairwise' panning, it proved to offer reasonably stable

Figure 7.20 Perceived location of phantom image versus interchannel level difference between side loudspeakers centred on 80° offset from front-centre, showing error bars. The forward loudspeaker is at 50° and the rear at 110°. It can be seen that the greatest uncertainty is in the middle of the range and that the image jumps rapidly from front to back. There is also more uncertainty towards the rear than the front. (After Theile and Plenge, 1977).

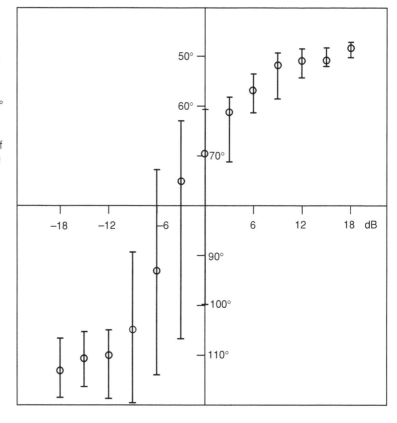

images for centre and off-centre listening positions, for moving and stationary sources, compared with some other more esoteric algorithms (see below). Front–back confusion was noticed in some cases, for sources panned behind the listener. His work used a standard five-channel loudspeaker arrangement. Martin *et al.*, on the other hand, experimented with pair-wise and other panning laws using a uniformly distributed circle of eight loudspeakers (Martin *et al.*, 1999), but they also found that the pair-wise approach provided the most focused phantom images of all. These results suggest that this panning method may have some useful applicability to surround sound recording, and it has the advantage of being easy to implement. Users simply have to be aware of what is and is not reliable.

7.2.2 VBAP

The amplitude panning concept has been extended to a general model that can be used with combinations of loudspeakers in arbitrary locations, known as vector based amplitude panning or VBAP (Pulkki, 1997). This approach enables amplitude differences between two or three loudspeakers to be used for the panning of sources, although it is not clear how the inaccuracy of amplitude panning between speakers to the sides of listeners may be overcome.

7.2.3 'Ambisonic' and other advanced panning laws

A number of variations of panning laws loosely based on Ambisonic principles have been attempted. These are primarily based on the need to optimise psychoacoustic localisation parameters according to low- and high-frequency models of human hearing. Gerzon based his proposals on the Makita theory of low-frequency localisation (Makita, 1962) (that is essentially Blumlein-style summing localisation based on interaural phase difference resulting from inter-speaker amplitude difference, represented by sine and cosine gain components forming a so-called 'velocity direction'), and on high-frequency energy-vector-based localisation based on the power differences between the channels. He rightly points out that the localisation position implied by these two vectors (velocity and energy) should be similar or the same for as many angles as possible, and that it rarely is with conventional amplitude panning except at the extremes of the image and in the centre. He therefore proposes a variety of psychoacoustically optimal panning laws for multiple speakers that can theoretically be extended to any number of speakers (Gerzon, 1992e). A three-channel version of this was described in Section 6.4.1. and Gerzon's own examples only extended up to four equally spaced loudspeakers. Some important features of these panning laws are:

- There is often output from multiple speakers in the array, rather than just two.
- They tend to exhibit negative gain components (out-of-phase signals) in some channels for some panning positions.
- The channel separation is quite poor.

A number of authors have shown how this type of panning could be extended to 5-channel layouts according to the standards of interest in this book. McKinnie (1997) proposed a 5-channel panning law based on similar principles, suitable for the standard loudspeaker angles given in ITU-R BS.775. It is shown in Figure 7.21.

Moorer found that the solutions cannot be optimal for the type of loudspeaker layout used in 5.1 surround systems as the angles between some loudspeakers exceed 90° and the layout does not involve equal spacing between the loudspeakers (Moorer, 1997). This rather knocks on the head the idea of using second order Ambisonics (see Section 4.8.3) for improving the directional accuracy of surround sound systems based on this loudspeaker layout, as Moorer shows that only the first spatial harmonic can be recreated successfully. Moorer (2000) plotted 4- and 5-channel panning laws based on these principles, pictured in Figure 7.22 (only half the circle is shown because the other side is symmetrical). He proposes that the standard ±30° angle for the front loudspeakers is too narrow for music, and that it gives rise to levels in the centre channel that are too high in many cases to obtain adequate L–R decorrelation, as well as giving rise to strong out-of-phase components. He suggests at least ±45° to avoid this problem. Furthermore, he suggests that the 4-channel law is better behaved with these particular constraints and might be more appropriate for surround panning.

Figure 7.21 Five-channel panning law based on Gerzon's psychoacoustic principles. (Courtesy of Douglas McKinnie).

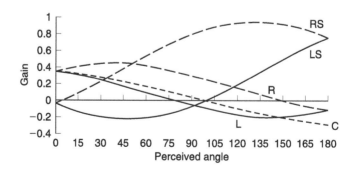

Figure 7.22 Two panning laws proposed by Moorer designed for optimum velocity and energy vector localisation with 2nd spatial harmonics constrained to zero. (a) Four-channel sound-field panning. The front speakers are placed at 30° angles left and right, and the rear speakers are at 110° left and right. (b) This shows an attempt to perform sound-field panning across five speakers where the front left and right are at 30° angles and the rear left and right are at 110° angles. Note that at 0°, the centre speaker is driven strongly out of phase. At 180°, the centre speaker is driven quite strongly, and the front left and right speakers are driven strongly out of phase. At low frequencies, the wavelengths are quite large and the adjacent positive and negative sound pressures will cancel out. At higher frequencies, their energies can be expected to sum in an RMS sense. (Courtesy of James A. Moorer).

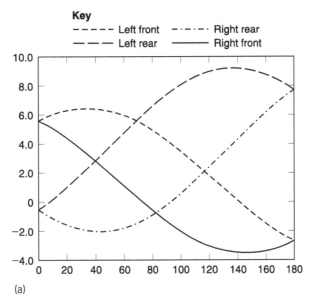

(a)

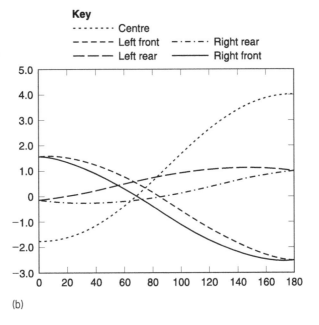

(b)

West (1999) tested a variety of panning laws including a hybrid surround-panning law based on Gerzon's optimal 3-channel panning law across the front loudspeakers (see Section 6.4.1) and conventional constant power panning between the remaining loudspeakers. He proceeded to develop an 'optimal' 5-channel panning law by combining elements of Gerzon's 3- and 4-channel versions, noting that Gerzon assumed equal spacing of loudspeakers and the modern layout does not, but that reasonable low and high frequency imaging was maintained. Again ±45° speaker locations were assumed in the front, for some reason. He then derived a solution based on Moorer's proposal to reduce second-order spatial harmonics to zero, again for the non-standard speaker locations.

All loudspeakers tended to give *some* output for *all* panning positions with these laws, and some of the outputs are out of phase with each other. This can give rise to unusual effects for off-centre listeners who might be close to out-of-phase loudspeakers, leading to severe skewing of the image away from the intended position, or phasiness. These laws tend only to be optimal for listeners in a relatively small 'hot spot'. In West's subjective tests he found that his 'optimal' 5-channel law was the most effective of those compared for the hot-spot listening position, for both stationary and moving pans, but that a simple constant power law was more stable for off-centre listening positions. His hybrid law (Gerzon in the front channels, constant power elsewhere) performed no better than either of these. The Moorer proposal performed less well in his tests.

The wish to avoid negative-going loudspeaker signals has inspired alternative proposals for panning laws based on similar principles, but having all positive components. An example is described by Dickins *et al.* (1999) of Lake DSP, termed 'non-negative least squares' (NNLS), which appears to sacrifice the requirement for congruent low and high frequency localisation vectors at the hot spot in favour of better image stability for off centre listeners. They do not show the panning coefficients needed for particular loudspeaker layouts. Their aim was to arrive at a solution that would work well for listeners over a large area such as an auditorium, and they analyse the success of their approach by looking at the magnitude of the energy vector (proposed by Gerzon as a measure of image focus for listeners at the hot spot, and of the stability of phantom images for off centre listeners). They show that the energy vector magnitude at the hot spot and away from it is greater than for either first or second order Ambisonics, for a regular array of six loudspeakers. To avoid large changes in image stability with moving pans, they 'smear' the panning coefficients so as to reduce the alteration in energy vector magnitude with panning angle. The effect of this is to make the phantom image slightly more diffuse in some positions than it would have been.

In Martin *et al.*'s subjective tests of image focus using different panning laws (and a non-standard loudspeaker array with loudspeakers every 45°), it was found that conventional pairwise constant-power panning provided the most focused images, followed by a relatively simple polarity-restricted cosine law (a non-negative gain approach, whereby values that would have gone negative are simply forced to zero) and second-order Ambisonics, followed by first-order Ambisonics and finally a novel law based on an emulation of a cardioid microphone polar pattern covering each 45° sector of the loudspeaker array. These tests were conducted at the hot spot only, and the authors subsequently concluded that the polarity-restricted cosine law appeared to create fewer unwanted side effects than the constant power law (such as changes in perceived distance to the source).

7.2.4 Head-related panning

Horbach of Studer has proposed alternative panning techniques based on Theile's 'association model' of stereo perception (Horbach, 1997; Horbach and Boone, 1999). This uses assumptions similar to those used for the Schoeps 'sphere' microphone described in Section 6.2.5, based on the idea that 'head-related' or pseudo-binaural signal differences should be created between the loudspeaker signals to create natural spatial images. It is proposed that this can work without transaural crosstalk cancelling, but that such crosstalk cancelling can be added to improve the full 3D effect for a limited range of listening positions.

In creating his panning laws, Horbach chose to emulate the HRTFs of a simple spherical head model that does not give rise to the high frequency peaks and troughs in frequency response typical of heads with pinnae. This is claimed to create a natural frequency response for loudspeaker listening, very similar to that which would arise from a sphere microphone used to pick up the same source. Sources can be panned outside the normal loudspeaker angle at the front by introducing a basic crosstalk cancelling signal into the opposite front loudspeaker (e.g. into the right when a signal is panned left). Front–back and centre channel panning are incorporated by conventional amplitude control means. He also proposes using a digital mixer to generate artificial echoes or reflections of the individual sources, routed to appropriate output channels, to simulate the natural acoustics of sources in real spaces, and to provide distance cues. This is derived from Theile's room-related balancing concepts.

7.2.5 Distance panning

Simulating distance in sound reproduction is a topic that has interested psychoacousticians for many years, but few distance panning features have been implemented in commercial sound mixers to date. This is probably because the signal processing capacity needed to simulate some of these cues has only recently become available. Now that digital mixers are commonplace it is possible that more sophisticated distance panpots may be implemented.

In Chapter 2 the basic differences that would be perceived between a distant source and a close source were summarised as:

- Quieter (extra distance travelled).
- Less high frequency content (air absorbtion).
- More reverberant (in reflective environment).
- Less difference between time of direct sound and first floor reflection.
- Attenuated ground reflection.

Figure 7.23 Distance simulation circuit (adapted from Gerzon, 1992a).

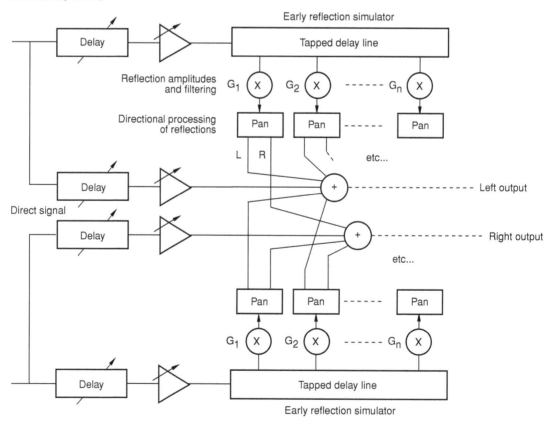

In addition we can add that for moving sources there is often a perceived Doppler shift (a shift in pitch either upwards or downwards) as a source moves quickly towards or away from the listener. Simulation of some or all of these effects can be used to create the impression of source distance in mixing, and Chowning was one of the first to describe the use of Doppler shift coupled with direct-to-reverberant ratio adjustment as a means of simulating moving sources (Chowning, 1971).

In addition to the coarse adjustment of direct-to-reverberant ratio in sound mixing, others have found that the timing structure of early reflections and reverberant energy can provide important cues relating to perceived distance. One such study was undertaken by Michelsen and Rubak (1997). Gerzon also concludes that the most important cues for distance simulation are provided by early reflections, and suggests a number of approaches for simulating such cues without the need for complete room modelling (Gerzon, 1992a). A simple distance simulation approach proposed by him is shown in Figure 7.23.

Figure 7.24 Changes in apparent sound source size with distance. Angle ϑ_1 (closer source) is greater than ϑ_2.

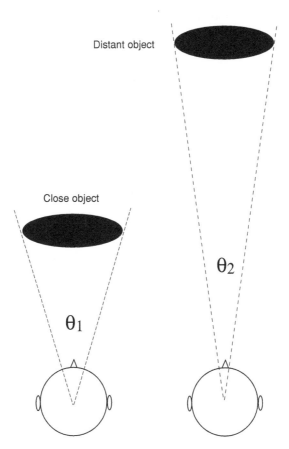

Distant object

Close object

θ_2

θ_1

Input

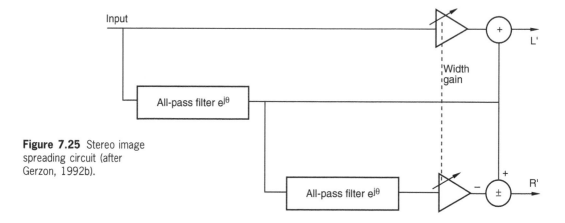

All-pass filter $e^{j\theta}$

Width gain

L'

All-pass filter $e^{j\theta}$

+

± R'

Figure 7.25 Stereo image spreading circuit (after Gerzon, 1992b).

Here the signal paths are assumed to be stereo, and the reflections are given different panning positions to the direct sound and to each other by means of a tapped delay line and a panning rotation matrix.

Gerzon also points out that the angular size of a sound source is different depending on its distance, as illustrated in Figure 7.24. A distant source will appear to be narrower than the same source closer to the listening position. Consequently he proposes a means of altering source size using various 'pseudo-stereo' processes that split the mono signal into two and filters at least one half using an all-pass filter that causes a frequency dependent phase difference between the two channels. This has the effect of causing the pan to swing between two locations depending upon frequency, resulting in the apparent spread of complex signals. One such circuit is shown in Figure 7.25 (Gerzon, 1992b).

7.3 Artificial reverberation and room simulation

Conventional studio reverberation units have tended to be based on a variety of delays and all-pass filters that create a collection of early reflections and diffuse reverberant tail with varying characteristics. Although many of the features of real acoustic spaces can be simulated with such units, they are not true 'room simulators' in that they do not mimic the reflection patterns of real sources at particular positions in a notional space. Such units also tend to be used for processing mono inputs and feeding decorrelated two-channel outputs, and all signals fed to one input are treated in the same way. A number of units are now appearing that can be used to feed 4- or 5-channel outputs rather than the usual two, for application in 5.1 systems.

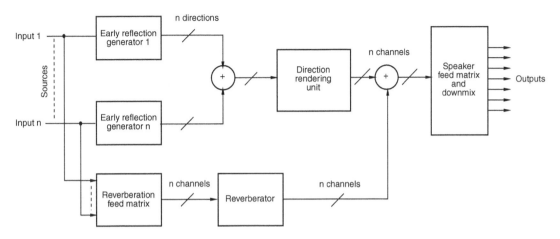

Figure 7.26 Rendering of room reflections in an artificial reverberation device (after Christensen and Lund, 1999).

In describing a recent approach to room simulation for multi-channel film and music applications, Christensen and Lund (1999) explain how a digital room simulator can be used to generate an appropriate reflection pattern for each sound source. The input sound sources can be panned to a particular position in the virtual space by the room simulator unit using a joystick or other control. The reflections are thereby different for each different source position, and consequently considered to be more natural and diverse. The resulting dry source and its reflected sound can be rendered for the appropriate loudspeaker reproduction format using either conventional amplitude panning, VBAP, Ambisonics or HRTF methods, as described above. A block diagram is shown in Figure 7.26.

7.4 Surround sound mixing aesthetics

7.4.1 To use or not to use the LFE channel?

The LFE channel was primarily designed for extra-low-frequency effects such as explosions and other loud LF noises in film mixing. The reason for this is that the reproduction level of film sound is normally calibrated to a certain sound pressure level (see Chapter 5), thereby limiting the maximum level of LF sound that can be reproduced from the main tracks when recorded at maximum level. The 'acoustic' headroom of the LFE channel is 10 dB greater than that of the other channels, owing to the method of reproduction alignment, enabling louder bass signals to be reproduced than would otherwise be possible. It was included in the consumer specification so that film mixes could be transposed directly to the home environment without the need for additional processing. In film reproduction the LFE channel feeds theatre subwoofers directly, whereas in consumer reproduction systems bass management is normally employed to enable both main channel LF content below a certain frequency *and* LFE effects to any subwoofer that might be installed, as shown in Chapter 4.

It bears repeating that, in mixing for any application, bass signals below 80–120 Hz do *not* have to be sent to the LFE channel unless one specifically wishes to do so for a particular effect purpose. It is quite normal for music mixing to use the main channels for the full signal bandwidth as would be the case with two-channel stereo. The LFE could be used exceptionally for emphasising dramatic effects, such as the bass drum wallops in the *Dies Irae* of the Verdi *Requiem*, for example, but would not be used as a rule. Also the LFE is typically discarded in any downmix to two-channel format such as might be executed by a Dolby Digital decoder for DVD, so any content will be lost.

7.4.2 What to do with the centre channel

The use of the centre channel, in music recording particularly, has been one of the most hotly debated topics in the move from two-channel to surround recording. Some engineers strongly protest that the centre channel is a distraction and a nuisance, and that they can manage very well without it, while others are equally strongly convinced of its merits. The psychoacoustical advantages of using a centre channel have been introduced in previous chapters, but it is indeed true that the need to generate suitable signals for this loudspeaker complicates panning laws and microphone techniques, as well as making down-conversion from surround to two-channel stereo more difficult. A possibly trivial but persuasive argument for the centre channel is that listeners with five loudspeakers will expect something to be coming out of all of them.

Those fighting against using the centre channel for music recording would like to be able to use recording techniques for the front left and right channels that are broadly similar to two-channel techniques, although possibly adjusting the direct to reverberant ratio of the signals to allow for the fact that the rear channels may contribute more of the reverberant energy than before. The ambience mikes or artificial reverberation that are often employed in classical recording are then used to feed the rear channels. Fold-down to conventional stereo can be a relatively simple matter of mixing some of the rear channels into the front, as described later. Some classical engineers find that simultaneous surround and 2-channel recordings of the same session are made easier by adopting 4-channel rather than 5-channel recording techniques, but this may be more a matter of familiarity than anything else. For many situations a separate mix and different microphones will be required for 2-channel and 5-channel versions of a recording. It is possible that a compromise could be reached by using a matrix such as that proposed by Michael Gerzon, described in Section 6.4, for deriving a centre channel from a 2-channel signal. In this way a conventional and compatible 2-channel microphone technique could be used for the front channels, being converted to a 3-channel version by the matrix.

In pop recording some engineers have claimed that using the centre channel makes central signals sound too focused and confined to the loudspeaker, but this is probably due to unsophisticated use of the format. In multitrack recording using panned mono sources, the panning law chosen to derive the feed to the centre channel will have an important effect on the psychoacoustic result. This is discussed in more detail in Section 7.2. Numerous studies have highlighted the timbral differences between real and phantom centre images, which leads to the conclusion that the equalisation of a source sent to a hard centre would ideally be different from that used on a source mixed to a phantom centre. Vocals, for example, panned so as only to emanate from the centre loudspeaker, may sound constricted spatially compared with a phantom image created between left and right loudspeakers, as the centre loudspeaker is a true source with a fixed location. Some 'bleed' into the left and right channels is sometimes considered desirable, in order to 'defocus' the image, or alternatively stereo reverberation can be used on the signal.

The technique of spreading mono panned sources into other channels is often referred to as a 'divergence' or 'focus' control, and can be extended to the surround channels as well, using a variety of different laws to split the energy between the channels. Holman advises against the indiscriminate use of divergence controls as they can cause sounds to be increasingly localised into the nearest loudspeaker for off-centre listeners. Gerzon's ideas relating to sound source spreading (see Section 7.2.5) could also be used in such circumstances to increase the perceived width of individual sources so that they do not appear to emanate from a point.

7.4.3 Dealing with surrounds

Surround channels in a 5.1 configuration, as already emphasised a number of times, are best reserved for mix components that one does not intend to be clearly or accurately localised, unless located specifically at loudspeaker positions. In film sound the concept of a surround 'loudspeaker position' is somewhat alien in any case, as there are usually numerous surround loudspeakers. In mixing music for consumer applications one may have the ability to treat the surround loudspeakers as more of a point source, although one cannot assume that they will be accurately sited by listeners. The difficulties of accurate panning in certain surround locations have been explained earlier, and while there may be some solutions that work better than others, the large angles between the surround speakers themselves and between surrounds and front make for unpredictable results at the best of times, particularly for off-centre listeners. This can be regarded as a hindrance to creativity, or it can be regarded as one of those disciplines that lead to greater and more structured artistic endeavour.

Although there are never hard and fast rules in sound mixing, it is probably fair to propose that sparing use of the surround channels will be best for the majority of applications, particularly for music purposes. While one wants to hear that there is a difference between surround and two-channel stereo, this should be handled in a subtle fashion in such a way that the listener feels more strongly enveloped and immersed in the production, rather than fatigued by dazzling and possibly over-stimulating effects. A lot depends on the application. Fast 'fly-bys' and other moving pans may be appropriate when tracking action in a movie picture, but are likely to be disconcerting in music balancing as there is not necessarily an accompanying visual element to 'explain' the action. Quite high levels of reverberation and other diffuse effects can be used in the surround channels before they become noticeable as separate sources, owing to the masking effect of higher level front channels. Diffuse ambient effects mixed to the surround channels should be in stereo wherever possible, with sufficient decorrelation between the channels to achieve a sense of spaciousness.

7.5 Upmixing and downmixing

Upmixing and downmixing (or up-conversion and down-conversion) are terms often used to describe the processes whereby one stereo format is converted to another, either from a small number of channels to a larger number or vice versa. Most common in current practice is the need to convert from 2-channel stereo to 4- or 5-channel surround (to create pseudo surround), and vice versa for creating a compatible 2-channel mix out of a surround master.

7.5.1 Synthesising surround from two-channel stereo

Just as true two-channel stereo cannot be synthesised from mono signals, so true surround cannot be synthesised from two-channel stereo. Nonetheless, just as attempts were made in the early days of commercial stereo to rework mono recordings as pseudo-stereo versions, using artificial 'spatialisation' algorithms, so many are interested in doing similar things in the early days of commercial surround.

There is an important difference, though, between the 1–2 channel situation and that arising when converting from 2–5 channels. It is principally that mono signals contain essentially no spatial information (except possibly distance cues), whereas two-channel signals contain substantial spatial information encoded in the differences between the two signals. Many algorithms that attempt to perform surround spatialisation of two-channel material do so by extracting some of the ambience signal contained in the difference information between the L and R channels and using it to drive the rear channels in some way, often with quite sophisticated directional steering to enhance the stereo separation of the rear channels. One can quite quickly see that this is the way that most analogue matrix surround decoders work (e.g. Dolby Surround), because the surround signal is usually encoded as a difference signal between left and right and has to be extracted to the rear loudspeakers. Sometimes a proportion of the front sound is placed in the rear channels to increase envelopment, with suitable delay and filtering to prevent front sounds being pulled towards the rear. Experiments by the author found that the level of signal extracted by such algorithms to the centre and rear channels was strongly related to the M and S (sum and difference) components of the 2-channel signal (Rumsey, 1998).

Matrix algorithms devised by Lexicon, Meridian and Circle Surround, such as described in Section 4.6, may be used for this purpose, although some are optimised better than others for dealing with two-channel material that has not previously been matrix encoded. Often a separate collection of settings is needed for upmixing unencoded two-channel stereo to surround than is used for decoding matrix encoded surround like Dolby Stereo. Meridian has adopted the so-called 'Trifield' principle proposed by Gerzon and others as a means of deriving a centre channel during the upmixing process (this is similar to the 2–3-channel conversion matrix described earlier in the book). Most of these devices are in fact consumer products designed for high-end home cinema systems. Dolby Surround decoders are not particularly successful for upmixing unencoded two-channel material as the surround signal remains mono and the centre is usually too strong, thereby narrowing the image quite considerably compared with the two-channel version.

There are also a number of (mainly consumer) algorithms used in some home cinema and surround systems that add 'effects' to conventional stereo in order to create a surround impression. Rather than extract existing components from the stereo sound to feed the rear channels they add reverberation on top of any already present, using effects called 'Hall' or 'Jazz Club' or some other such description. These are not recommended for professional use as they alter the acoustic characteristics of the original recording quite strongly rather than basing the surround effect on the inherent spatial qualities of the original recording.

Figure 7.27 Subjective quality differences between original two-channel and upmixed five-channel renderings of stereo programme items for four anonymous processors. A ten point scale was used. Positive diffgrades are equivalent to a judgement of greater quality. (a) Front image quality grades show clear agreement between subjects about the reductions in front imaging quality. (b) Spatial impression grades demonstrate much less consistency among subjects.

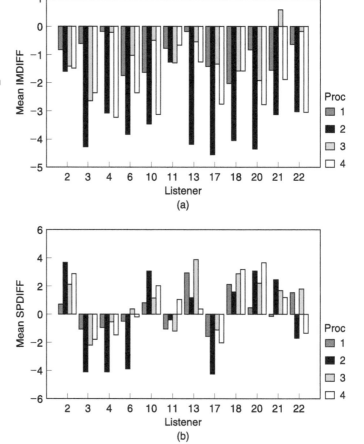

Subjective experiments carried out by the author on a range of such upmixing algorithms found that the majority of two-channel material suffered from a degradation of the front image quality when converted to five-channel reproduction (Rumsey, 1999). This either took the form of a narrower image, a change in perceived depth or a loss of focus. On the other hand, the overall spatial impression was often improved, although listeners differed quite strongly in their liking of the spatial impression created (some claiming it sounded artificial or phasey). The results are summarised for four anonymous processors in Figure 7.27. Interestingly, the majority of expert listeners who took part in these blind tests tended to prefer the two-channel version to the five-channel version tests, suggesting that synthetic five-channel surround is considered less acceptable by experts than good two-channel sound. Familiarity with the spatial effect created by some algorithms led to less dissatisfaction and settings could often be found that created a pleasing spatial effect for a range of programme material, with enhanced envelopment and only slight reduction in the front image quality.

7.5.2 Downward compatibility of multichannel mixes

A tricky problem in the increasingly multiformat world of two-channel and multichannel mixes is the downward compatibility of multichannel mixes. A number of options exist here. One can undertake a completely separate mix for each format, which may be the most aesthetically satisfactory approach, optimising the result for the format concerned in each case. Unfortunately this can be extremely time consuming, and this has led to the need for semi-automatic or automatic downmixing of multichannel mixes. Such techniques are useful in consumer equipment and in broadcasting environments where one needs to accommodate listeners that do not have surround sound replay and where there may not be a separate two-channel mix available. The problem is similar to that of stereo-mono compatibility although somewhat more complicated because there are more possibilities.

A number of engineers have commented that the total amount of reverberant sound in multichannel mixes can be different to that in two-channel mixes. This is partly because the spatial separation of the loudspeakers enables one to concentrate on the front image separately from the all-round reverberation, whereas in two-channel stereo all the reverberation comes from the front. Directional masking effects also change the perception of direct-to-reverberant ratio in the two formats. Simply adding the rear channels into the front channels at equal gain, and mixing the centre into left and right, may create an over-reverberant two-channel balance and one that has too narrow a front image. Consequently some control is required over the downmix coefficients and possibly the phase relationships between the channels, for optimal control over the two-channel result. Even with this, it may be that a downmixed two-channel version is never as satisfactory as a dedicated two-channel mix, especially for critical material such as classical music.

7.5.3 ITU downmix parameters

Downmix equations are given in ITU-R BS.775, intended principally for broadcasting applications where a 'compatible' two-channel version of a five-channel programme needs to be created. These are relatively basic approaches to mixing the LS and RS channels into L and R respectively, and the centre equally into front left and right, all at –3 dB with respect to the gain of the front channels. Recognising that this may not be appropriate for all programme material the recommendation allows for alternative coefficients of 0 dB and –6 dB to be used. Formulae for other format conversions are also given.

Experiments conducted at the BBC Research Department suggested that there was little consistency among listeners concerning the most suitable coefficients for different types of surround programme material, with listeners preferring widely differing amounts of surround channel mixed into the front. It is possible that this was due to listeners having control over the downmix themselves, and that in cases where there was little energy in the surround channels a wide range of settings might have been considered acceptable. Averaged across all programme types a setting of between –3 and –6 dB appeared to be preferred, albeit with a wide variance.

7.5.4 Dolby Digital downmix control

Dolby Digital decoders provide a range of downmixing options from five-channel down to matrixed L_T/R_T (matrix encoded LCRS Dolby Surround) or to a two-channel version (L_0/R_0). The downmix can also take into account whether the different speakers are large or small, whether a subwoofer is used and various bass management options. The most useful feature of this system is that the downmix coefficients can be varied by the originator of the programme at the post-production or mastering stage and included as side information in the Dolby Digital data stream.

In this way the downmix can be optimised for the current programme conditions and does not have to stay the same throughout the programme. Listeners can choose to ignore the producer's downmix control if they choose, creating a custom version that they prefer.

Downmixing from 5-channel Dolby Digital to matrixed Dolby Surround is accomplished by mixing the centre channel into L and R at –3 dB. LS and RS are combined with an overall gain of –3 dB and the sum mixed out of phase into L and R with a bandwidth of 100 kHz to 7 kHz. The 90° phase shift employed in LS and RS in the Dolby Digital encoder (see Chapter 4) avoids the need for it to be created in the downmix decoder (which is difficult in DSP). Downmixing to a two-channel L_0/R_0 version is done in a similar way to the ITU recommendation described above, with alternative gains for the centre and surround channels of:

C: –3, –4.5 or –6 dB;

LS and RS: –3, –6 or –∞ dB (mixed into L and R with no phase modification);

LFE: not included in downmix.

7.5.5 Downmix control in DVD-Audio

DVD-Audio is expected to employ so-called SMART (System Managed Audio Resource Technology) downmixing to provide compatible two-channel outputs from five-channel PCM mixes stored on the disk. The option also exists for a separate two-channel mix to be stored on the disk alongside the five-channel version, provided there is space. SMART downmixing allows the gain, panning and phase of the centre and surround downmix to be indicated during mastering, with coefficients stored on the disk to control the decoder. This can be done on a track-by-track basis.

If the disk is MLP encoded (see Section 4.7.5) a two-channel downmix can be stored on the disk that takes up very little additional space compared with the surround version, enabling a separate producer-controlled mix to be stored, although probably still based upon some combination of the centre and surround channels with the front channels.

7.5.6 Gerzon compatibility matrix

It would be surprising if Michael Gerzon had not said anything about the matter of downmixing, and indeed he did tackle the issue in a comprehensive attempt to propose means of converting between any combination of stereo/surround formats (Gerzon, 1992c, d). This formed part of his massive contribution (some seven lengthy papers) to the San Francisco AES Convention in 1992. In this paper he proposes that, in order to preserve stereo width and make the downmix hierarchically compatible with other formats, an alternative downmix formula from 5–2 channels should be used (Gerzon's notation adapted to that used in this book):

$L_0 = 0.8536L + 0.5C - 0.1464R + 0.3536k(LS + RS) + 0.3536k2(LS - RS)$

$R_0 = -0.1464L + 0.5C + 0.8536R + 0.3536k(LS + RS) - 0.3536k2(LS - RS)$

Where k = between 0.5 and 0.7071 (–6 and –3 dB) and $k2$ = between $1.4142k$ and 1.4142 (–3 to +3 dB).

Here he assumes that the 3/2 stereo approach is used whereby three-channel front stereo is assumed, plus separate surround signals. The result of his matrix is that the front stereo image in the two-channel version is given increased width compared with the ITU downmix proposal (which would narrow the image somewhat), and that the rear difference gain component k_2 has the effect of making rear sounds reproduce somewhat wider than front sounds. He suggests that this would be generally desirable because rear sounds are generally 'atmosphere' and the increased width would improve the 'spatial' quality of such atmosphere and help separate it from front stage sounds. Based on the above equation he proposes that values of $k = 0.5$ and k_2 = 1.1314 work quite well, giving a folded-down rear stage about 4 dB wider than that of the front stage and with the rear channels between 3.5 and 6 dB lower in level than the front.

7.5.7 Logic 7 downmixing

Griesinger (2000) has described the basic principle of a downmixing approach that is used in Lexicon's Logic 7 surround algorithms. This matrixing technique is designed primarily to create a two-channel signal that can subsequently be de-matrixed back to surround, but it can also be used to create a plausible two-channel downmix of five-channel material. Although the full implementation of the algorithm is slightly more sophisticated than this, the broad principle is based on the following:

$L_0 = L + 0.707C + 0.9LS - 0.38RS$

$R_0 = R + 0.707C + 0.9RS - 0.38LS$

In other words, the centre is mixed into left and right at –3 dB, and the surrounds are each mixed into the same side's front channel with a degree of anti-phase crossfeed into the opposite channel. (To be correct, there is also a 90° phase shift in the rear signals in the matrix, but a basic downmix implementation can be made without this.)

If the average of the absolute value of LS and RS (whichever is greater) is less than the average value of L, C, or R, (whichever is greater) by 3 dB, the gain of the mix of LS and RS is reduced. The gain reduction starts when this difference is 3 dB, and is complete (at –3 dB) when the difference is 6 dB or greater. The purpose of this gain reduction is to make the downmix compatible with the European standard, which specifies the –3 dB attenuation, while maintaining full gain for strong signals in the rear.

The relative phase of LS and RS is detected. When they are in phase, the gain of the LS and RS mix is reduced by up to 3 dB. Thus if LS and RS are decorrelated or separate, full gain is used. When they have a mono component, the gain of the mix is reduced so as to preserve the total energy in the output. The purpose of this gain reduction is to preserve total energy in the encoder, and to make it compatible with the Dolby Film encoder, which has a 3 dB gain reduction.

References

Bruck, J. (1997). The KFM 360 Surround – a purist approach. Presented at *103rd AES Convention, New York, 26–29 Sept*. Preprint 4637. Audio Engineering Society.

Chowning, J. (1971). The simulation of moving sound sources. *J. Audio Eng. Soc.*, **19**, 1, pp. 2–5.

Christensen, K.-B. and Lund, T. (1999). Room simulation for multichannel film and music. Presented at *AES 107th Convention, New York, 24–27 September*, Preprint 4933. Audio Engineering Society.

Dickins, G., Flax, M., McKeag, A. and McGrath, D. (1999). Optimal 3D speaker panning. In *Proceedings of the AES 16th International Conference*, pp. 421–426. Audio Engineering Society.

Fukada, A., Tsujimoto, K. and Akita, S. (1997). Microphone techniques for ambient sound on a music recording. Presented at *103rd AES Convention, New York, 26–29 Sept*. Preprint 4540. Audio Engineering Society.

Gerzon, M. (1992a). Design of distance panpots. Presented at *92nd AES Convention, Vienna*. Preprint 3308. Audio Engineering Society.

Gerzon, M. (1992b). Signal processing for simulating realistic stereo images. Presented at *93rd AES Convention, San Francisco, 1–4 October*. Preprint 3424. Audio Engineering Society.

Gerzon, M. (1992c). Optimum reproduction matrices for multispeaker stereo. *J. Audio Eng. Soc.*, **40**, 7/8, pp. 571–589.

Gerzon, M. (1992d). Compatibility of and conversion between multi-speaker systems. Presented at *93rd AES Convention, San Francisco, 1–4 October*. Preprint 3405. Audio Engineering Society.

Gerzon, M. (1992e). Panpot laws for multispeaker stereo. Presented at *92nd AES Convention, Vienna*. Preprint 3309. Audio Engineering Society.

Griesinger, D. (2000). Personal communication.

Hermann, U. and Henkels, V. (1998). Vergleich von 5 surround-mikrofonverfahren. In *Proceedings of the 20th Tonmeistertagung, Karlsruhe*, pp. 508–517. VDT.

Holman, T. (1999). *5.1 Surround Sound: Up and Running*. Focal Press, Oxford and Boston.

Horbach, U. (1997). New techniques for the production of multichannel sound. Presented at *103rd AES Convention, New York, 26–29 September*. Preprint 4625.

Horbach, U. and Boone, M. (1999). Future transmission and rendering formats for multichannel sound. In *Proceedings of the AES 16th International Conference*, pp. 409–418. Audio Engineering Society.

Klepko, J. (1997). 5-channel microphone array with binaural head for multichannel reproduction. Presented at *103rd AES Convention, New York, 26–29 September*. Preprint 4541. Audio Engineering Society.

Makita, Y. (1962) On the directional localisation of sound in the stereophonic sound field. *EBU Review*, part A, no. 73, pp. 102–108.

Martin, G., Woszczyk, W., Corey, J. and Quesnel, R. (1999). Controlling phantom image focus in a multichannel reproduction system. Presented at *107th AES Convention, New York, 24–27 September*. Preprint 4996. Audio Engineering Society.

Mason, R. and Rumsey, F. (1999). An investigation of microphone techniques for ambient sound in surround sound recordings. Presented at *106th AES Convention, Munich, Germany, 8–11 May 1999*. Audio Engineering Society.

McKinnie, D. (1997). Personal communication.

Michelsen, J. and Rubak, P. (1997). Parameters of distance perception in stereo loudspeaker scenario. Presented at *102nd AES Convention, Munich*. Preprint 4472. Audio Engineering Society.

Mitchell, D. (1999). Tracking for 5.1. *Audio Media*, November, pp. 100–105.

Mora, A. and Jacques, N. (1998). True space recording system. Measuring angular distortion. *www.multimania.com/tsrs*

Moorer, J. (1997). Towards a rational basis for multichannel music recording. Presented at *103rd AES Convention, New York, 26–29 September*. Audio Engineering Society.

Moorer, J. (2000). Personal communication.

Pulkki, V. (1997). Virtual sound source positioning using vector base amplitude panning. *J. Audio Eng. Soc.*, **45**, 6, pp. 456–466.

Rumsey, F. (1998). Synthesised multichannel signal levels versus the M-S ratios of 2-channel programme items. Presented at *AES 104th*

Convention, Amsterdam, 16–19 May. Preprint 4653. Audio Engineering Society.

Rumsey, F. (1999). Controlled subjective assessments of 2–5 channel surround sound processing algorithms. *J. Audio Eng. Soc.*, **47**, 7/8, pp. 563–582.

Theile, G. and Plenge, G. (1977). Localisation of lateral phantom images. *J. Audio Eng. Soc.*, **25**, pp. 196–200.

Theile, G. (2000). Multichannel natural music recording based on psychoacoustic principles. Presented at *108th AES Convention, Paris, 19–22 Feb.* Preprint 5156. Audio Engineering Society.

West, J. (1999). Masters thesis. University of Miami, Florida.

Williams, M. and Le Dû, G. (1999). Microphone array analysis for multi-channel sound recording. Presented at *107th AES Convention, New York, 24–27 Sept.* Preprint 4997. Audio Engineering Society.

Williams, M. and Le Dû, G. (2000). Multichannel microphone array design. Presented at *108th AES Convention, Paris, 19–22 Feb.* Preprint 5157. Audio Engineering Society.

Index